MW00748450

# The 21ˢᵗ Century Environmental Revolution:

## A Comprehensive Strategy for Conservation, Global Warming, and the Environment

/

## The Fourth Wave

*First Edition*

**Book I**
**Waves of the Future Series**

# By Mark C. Henderson

**Waves of the Future**
http://wavesofthefuture.net

Published
by
Waves of the Future
*First Printing*

Copyright © 2008 Waves of the Future.  All rights reserved.
Cover Design Copyright © 2008 Waves of the Future.
Cover images: NASA

Cataloguing in Publication

Henderson, Mark C.
   The 21st century environmental revolution : a comprehensive strategy for conservation, global warming, and the environment : the fourth wave / by Mark C. Henderson. -- 1st ed.


(Waves of the future series)
Includes bibliographical references.
ISBN 978-0-9809989-0-0


   1. Environmental sciences.  2. Environmental protection.  3. Environmental economics.  I. Title.  II. Series.

GE105.H45 2008                     333.7                     C2008-
902931-3
Library and Archives Canada

Waves of the FutureSeries
http://wavesofthefuture.net

# Table of Contents

# *Acknowledgment and Foreword*

Firstly, I would like to express my gratitude to Pierre Bédard (editor) who has supported my work and helped make this book a reality.

This work is international in scope. One of the main difficulties in writing this book arose from the need to reach out to a broad audience. Political changes do not happen without the support of a large number of people.

On the one hand, the challenge was to make issues, concepts, and solutions readily understandable to the vast number of those who have the real power, the voters. On the other hand, the solutions presented needed to be based on sound scientific and economic grounds in order to have any hope of success. Hence, there was a need to also make a solid case to the academic establishment, and this, without getting overly technical and putting off voters.

I have tried to achieve these goals by aiming most discussions at the broadest possible audience, covering both basic and more advanced concepts in the most accessible ways.

# Book II
## of the
## *Waves of the Future Series*
## is already in the works.

**Check the publisher's website for release dates and places to buy.**

**http://wavesofthefuture.net/**

# *Introduction*

Finally, we are seeing some action on the environment. Unfortunately, it comes after disaster has hit and people have died. Hurricane Katrina and the late summer 2005 New Orleans flooding was a wakeup call for many. Today, governments are increasingly moving towards curtailing greenhouse gases in an attempt to slow down global warming. If we are going to wait until disaster hits before acting, then people are going to die each time. We need to be smarter.

Environmental issues are much bigger than global warming alone. The number of toxic chemicals and carcinogens in our own body tissues is simply shocking. World resources such as the oceans' fisheries are beginning to be degraded and eroded by toxic compounds. The Kyoto Accord on greenhouse gases is much too small and narrow in focus to address today's environmental challenges with any effectiveness. We need to look at much more profound changes and contemplate a much larger scale of action. We need to go to the next step. This is what this book is about.

Decades ago, Alvin Toffler wrote about three waves of change that swept over the world and transformed societies on a massive scale in the process. This is the beginning of a new one: the Fourth Wave.

Most of us woke up this morning assuming that the sun would rise. Beyond that little can actually be assumed. We know far less than we think about what really lies ahead of us. A decade ago, global warming was considered by many but a theory, the fancy of some scientist's imagination. Now, it is all but a certainty, one that we will have to live with for decades. That is, assuming we are eventually able to reverse course.

Many view the 20th century as a time of progress. There has certainly been a staggering number of technological advances during that period. From a social and economic point of view, many positive developments have also occurred. However, we have failed to address several of the problems that plagued societies during that period. Many could have been solved a decade or more ago given enough commitment.

We have made some progress in terms of world peace, but we have also developed weapons of mass destruction capable of wiping out life on the entire planet multiple times, as if one was not enough. Not enough headway has been made in the fight against unemployment, social disparities, and world poverty given the vast increase in wealth we saw in the second half of the century. Even the richest countries in the world have failed at eradicating homelessness and dire poverty on their own streets.

In a relatively short period of industrialization, we have succeeded in producing a tremendous amount of pollution, some of which reaching just about everywhere including Antarctica, some of which lasting for extremely long periods of time.

There is a problem with the current economic system, at least with the way it is set up. To create jobs, we need to raise consumption levels. However, this also means increasing pollution and using up resources. China and India are viewed as vast consumer markets with a huge potential for job creation. But few speak of the effects that such massive consumption would have on the environment and the depletion of non-renewable resources.

Two things are increasingly becoming clear. Firstly, we cannot continue down the path we are currently on. Our impact on the planet is simply too devastating. Fundamental changes to the economic system have to occur. Secondly, our destructive powers have significantly grown in the last 100 years. That trend is accelerating as world production increases and mass markets are only getting better. At the beginning of the 20th century, we did not have the technology to cause a lot of damage. Today, we have our pick of ways to destroy ourselves and the planet: weapons of mass destruction, pollution, the total depletion of essential non-renewable resources, etc.

## Looking Ahead

There is much currently going on in the world, a lot of which will determine the kind of future we will move into. So far, world development has largely been determined by technological waves. However, social factors have also played an important part. One may think, for example, of the way democracy has fashioned modern societies or the impact of the human rights movement. In the current geopolitical chaos, the world may make a turn for the better or the worse. The stakes are high.

We have reached a point in history where things cannot continue to go on the way they have in the past. There can be good times ahead even for those currently at the top, but these will come at a price. History has shown many an empire crumbling because of their inability to shift thinking and challenge fundamental assumptions when the time had come. In the years ahead, countries will have to decide whether to maintain the status quo and move towards decline and disaster or to challenge their assumptions and trade a little pain now for a much brighter future.

There is generally a lot of resistance to new ideas, and things usually change very slowly. This is often for the better, preventing society from hastily going down disastrous paths. However, we do not have the luxury of time on many environmental issues—unless, that is, we don't care about our children, grandchildren, and future generations. This is the time for bold actions and audacity, not for endless denials and delaying tactics.

Many of us are starting to feel that the political agenda has plateaued. Few new and creative ideas have occurred in the last 20 years. The Fourth Wave will bring many fresh ideas and challenge us into reassessing the past and defining a new agenda for the future.

## The Fourth Wave

*The 21st Century Environmental Revolution* is the initial book of a series dealing with contemporary issues. Its first part provides a historical perspective on current social and economic issues by going back in time through Toffler's first three waves of change. It makes the case that a fourth one is about to hit and transform society and the world in which we live faster and more fundamentally than anyone expected.

The second part of the book deals with the usual spectrum of environmental issues but also the question of non-renewable resource use.

The third part focuses on what lies ahead of us: an environmental revolution early in this century. It lays out a market-based strategy that would make possible environmental change on a large scale and over a relatively short period of time.

As this book targets a broad audience, some readers will have difficulty with a few sections which are more complex. Feel free to jump to the following heading if need be. Others will find some parts too easy or general in scope. If your knowledge of the environment is good, just read selectively through those. The new strategy begins at chapter 6.

This book puts together a new environmental strategy which is capable of being implemented on a large scale. It is a piece of research written in easy to understand language and in a format interesting for everybody. As a result, more people can be reached and greater political support gained.

The environmental approach proposed in this book could take us out of the destructiveness of the $20^{th}$ century into a bright new future. It would bring up new vistas of hope and could transform the world in ways nobody could have anticipated only a short time ago.

Welcome to the Fourth Wave.

# 1. Non-Technological Revolutions

## The Sexual Revolution

Unlike Toffler's first three waves, which arose from technological change, the fourth one is socially based. As such, it has been foreseen by very few and will come unexpectedly. The sexual liberation in the 1960s was not given rise by technology. Yet, we talk of a revolution, one that has fundamentally changed our lives. Who would have dreamed in the late 1950s that a decade later people would run around naked in a farmer's field, adorned with flowers and listening to rock music while talking about taking over the world with peace and love? That was Woodstock, August 16, 1969.

The new technology of the Industrial Revolution could not account for that. A change in morality was the determining factor. It came as a result of a combination of things. At a time when industrialization had peaked and was bearing fruits for everybody, there was a sense of freedom, of wealth. For the first time in history, we were breaking the chains of economic enslavement; we could afford to design a world of beauty. The times were ripe for change.

This powerful mix of events and forces turned sex—which had been dirty and sinful up until then—into an expression of freedom and beauty. Although birth control technology had been around for a long time, the sexual revolution only happened when its time had come: after a number of changes in values and thinking, as well as social organization—for example, the separation of Church and state. It is beyond the scope of this book to examine those issues in any detail. Suffice it to say that if technology had been the determinant factor in the sexual revolution, the latter would have happened in the 18th century, when condoms were first invented.

Major changes and revolutions can happen and have a powerful effect on society without being technological in nature. They also tend to come more unexpectedly because human factors are less tangible and predictable than technological trends. Sometimes, a single idea can spark a revolution.

If anything, we should be much more afraid of ideas than technology. Although the Fourth Wave is about positive change, its power will have to be reckoned with.

## The Right Recipe for a Revolution

One of the key ingredients in a revolution is ideas, as already discussed. Another one is popular support. Current political commitment levels can take us some distance but are often dependent on people falling ill or dying and will certainly fall short of creating a massive wave of change on their own. Fortunately, revolutions are often the result of a combination of elements.

Timing is very important. Winning the right of freedom of speech made the ground fertile for many changes. Before it happened, social movements, such as the sexual liberation, gender equity, and unionizing, were mere possibilities. After it did, they became virtual certainties. So, what appears to be a choice at first may not be one at all but a powerful and unstoppable force for change.

Fundamental legitimacy is a key element in evaluating the inevitability of proactive revolutions, in determining whether political leaders around the world can simply dismiss forthcoming ideas or will find themselves waking up in cold sweat at night, trying to think ahead of the tidal wave of change looming on the horizon. The democracy, freedom of speech, and gender equity movements overcame incredible odds primarily because of their rightfulness.

Given the right timing and fundamental legitimacy, non-technological revolutions can happen unexpectedly and be very powerful. We are at the beginning of the new millennium. There are few new items on the agenda, just more of the same old solutions to the same old problems. The Fourth Wave will change that.

## Entering the Third Millennium

In his 1980 book, *The Third Wave*, Alvin Toffler talks about three massive waves of change bringing about a total transformation in the way people worked, lived, and related to each other. He (1980) describes the Third Wave as "a powerful tide ... surging across much of the world ... creating a new, often bizarre, environment in which to work, play, marry, raise children, or retire" (p. 17). His work inspired many and became a cult classic in the early 1980s.

In his analysis, Toffler describes waves of change sweeping across the globe at different pace in different areas, leaving some early societies untouched but propelling others into the future. They overtook each other, bringing about transformation in their passage. They also clashed together, creating the disruptions associated with the birth pains of new eras.

The tides of change Toffler (1980) talks about began with the First Wave, or the agricultural revolution that took place during the Neolithic period of human history. While the First Wave lasted for thousands of years, the second one—the Industrial Revolution—began around 1750 CE and has already seen its dominance overtaken by the next one. He expected that we would fully bear witness to the Third Wave—the Information Age—and see it completed within possibly only decades (p. 26).

Although some of his predictions have not come true, many did to a greater extent than he even anticipated. The Internet and the computer age are good examples of this. Even he, did not forecast the extent to which they would totally pervade our lives from the inside out. Toffler argued that change was accelerating. This phenomenon will have a lot of relevance as the Fourth Wave sweeps across the planet in the years to come.

## The Master Plan for the New Millennium

We are now standing at the threshold of the Fourth Wave, and, if I may rephrase Toffler's (1980) words about the Third Wave, "We, who happen to share the planet at this explosive moment, will ... feel the full impact of ... [this new wave] in our own lifetimes" (p. 26).

The new millennium is no longer around the corner; it is here. What is our plan for the future? What would we like to achieve in the next century? Do you recall hearing of any plan for the long-term future, or even this decade? The extent of planning for this millennium seems to have limited itself to coping with emergencies as they come up, not before they do, and with the predictable death tolls. Sure, there has been a certain amount of noise in terms of what we want, but little of significance.

At this point, we barely manage to cope with making laws to control new technology as it appears and evolves. The Internet and its new wave of crime is a case in point.

### Our Legacy

What kind of planet will our children inherit from us? Will it be better or worse than the one that was bequeathed to us? What will be our final legacy?

There have been bits and pieces of legislation made with respect to the environment but nothing significant, judging from the wastefulness in the packaging industry, the increase of contaminants everywhere, the intensive use of chemicals in the agricultural industry, or the rise of environmental polluters like low-gas-mileage vehicles. Our efforts in addressing global warming problems have had so far limited support and results.

The ozone layer depletion has turned a boon, the sun, into a scourge to the point where health authorities now recommend the use of sun-blocker creams at all times. While this goes on, nutritionists point out that these precautions can lead to vitamin D deficiencies and recommend spending some time in the sun without protection!

Fresh water is now considered a health hazard to the point where in many places it is even dangerous to swim in, not to speak of drinking it! No matter where you go, mercury and other pollutants are found in increasing levels in all fish species. Yet, we keep spewing mercury into the atmosphere through a number of industrial processes.

Our own body tissues and those of our children are laced with hundreds of toxic compounds and carcinogens. The milk we feed our kids, the water they drink, the food they eat are full of contaminants. While some do not pose dire problems now, we know that they cumulate over time and progressively do permanent damage to the environment.

## Non-Renewable Resources

Few speak of the depletion of non-renewable natural resources. Everybody has heard about saving trees and global warming, but how much of the earth's mineral resources are we going to leave future generations? The metals, upon which much of our lives and lifestyles depend, are not renewable. Once used up, they do not grow back. We are depleting them at an ever accelerating rate.

We need enough resources not only for our grandchildren but also for theirs and the generations to come. The life expectancy of the earth is about 5 billion years (Contemporary World Atlas, 1988, p. 6A). That is a long, long time for which people will need metals.

Currently, the only planning with respect to non-renewable resources is a conspiracy of silence: take as much as you can, and don't ask any questions or raise the issue. What we are doing at this point in time with respect to non-renewable resources can only be characterized as totally reckless. We have to do better.

Along with the pressing need to address the problem of environmental contamination, there is a dire urgency to stop the plundering of non-renewable resources.

## Looking to the Fourth Wave Future

We can continue to react to crises and then deal with the aftermath of disasters. We can continue to march backward into the future, or we can make choices to bring about the world that we want.

The Fourth Wave will put a stop to the chronic and wanton destruction of the planet and to the reckless wasting and plundering of non-renewable resources. It will fundamentally and radically change the way we relate to the world we live in.

# 2. Toffler's Waves of Change

## The Pre-First Wave

How would you define our society? Perhaps, looking back at where we come from will help in answering this question. As this book is partly rooted in Alvin Toffler's works, his own descriptions and definitions will be used.

Pre-first-wave societies go back over 10,000 years ago. Toffler describes them as small bands of nomads living off the land by fishing, hunting, and gathering wild plants. They still exist today in some parts of the world—for example, some tribes in the tropical forests of the Amazon and Papua New Guinea—but, generally, their way of life gradually came to an end when the First Wave—the agricultural revolution—began taking hold about 10,000 years ago (p. 29).

### Pre-Agricultural Life

The pre-agricultural era is generally defined by anthropologists as the first part of the Stone Age—the Old Stone Age or the Paleolithic (500,000 BCE to about 10,000 BCE). Early Stone Age people were nomads who lived off the land. They constantly moved in pursuit of food. They followed the herds of animals they hunted and moved away from an area once it was depleted of game or the plants (fruit, berries, roots, leafy greens, nuts, grains, etc.) they ate (Beers, 1986, p. 21). Their diet changed depending on food availability and season of the year.

Their social structure was generally simple. Their way of production was not very effective. Because they migrated regularly, everything they owned had to be carried with them. As a result, they could not accumulate food surpluses and only produced what was needed. When hard times hit—as a result of droughts or other natural causes —they simply starved.

## The First Wave: The Agricultural Revolution

Toffler describes the First Wave as having begun with and been driven by the agricultural revolution. He locates it in time between 8000 BCE and 1650-1750 CE. The first part of this period—till about 3500 BCE—took place during the New Stone Age, or the Neolithic. It was characterized by the birth and growth of agriculture.

During that period, people gradually moved from hunting and gathering to herding and agriculture. That span of time saw the domestication of wild animals, for example, dogs, sheep, and goats. Vegetables and grains were grown and harvested. Different regions of the world saw different types of food being cultivated, depending on the suitability of soils and climates. Rice and yams were grown in Asia. Wheat, oats, and barley were staples of the Middle East and Africa. South America's traditional crops were maize and beans (Beers, 1986, p. 22).

The transformations that the agricultural revolution brought to the Stone Age way of life and the implications it was to have for the future of society were staggering. The adoption of agricultural practices led to greater and more secure supplies of meat, grains, and other foods. The risks of starving from having a bad hunting season were lessened, and people no longer had to move constantly. Reserves existed in the form of herd stocks and grain stores.

Cultivation required the development of a number of new tools and the building of facilities for the storage of seeds. It necessitated land, which not only had to be fertile but also needed to be made ready or suitable for food growing; it had to be freed of trees and root systems, leveled and drained, tilled, etc. That led to the abandonment of the nomadic way of life. As facilities and cleared land could not be moved around and involved a lot of work, people began settling down around them.

## *The Passing on of Wealth*

A significant economic aspect of agriculture was that wealth or capital could be built over time and passed on to the next generation. Stores of grain, land, equipment, and facilities could be accumulated over the years and bequeathed to the following generation. Likewise, the main stock of a herd could also be passed on to one's children.

This meant that not only safety but also wealth and power could accumulate over the generations. As a result, certain families, clans, and tribes emerged as wealthier and more powerful. The economic surplus and the ability to pass it on laid the groundwork for a social hierarchy to emerge.

Toffler (1980) talks of a generally parallel evolution in all major civilizations—be it in Europe, Asia, or Latin America. Land had become the basis of society. Life—economic, cultural, familial, and political—revolved around it. The village became the center of the social organization. They all saw the emergence of basic work specialization (division of labor). Rigid and authoritarian social classes (the nobility, priesthood, military, peasantry, etc.) began to appear. Rank was determined by birth, not merit (pp. 37-38). Neolithic economies were decentralized, and communities were for the most part self-sufficient.

## *The Rise of the City*

The other important development that happened as a result of increased farming and agricultural surpluses was the rise of the city. Cities first appeared around 6000 BCE, but the actual urban revolution that marked the beginning of civilization occurred around 3500 BCE. The Tigris and Euphrates rivers in the Middle East, the Nile in North Africa, and the Indus and Yellow rivers in South Asia and Asia were the cradles that gave birth to civilization.

Cities were largely oversized agricultural settlements at first. With the growth of food surpluses, they were able to support a larger non-agricultural class. As the need for tools and technology increased, a new urban class grew, the artisans. They were the forerunners of modern tradespeople.

While at the beginning of this new millennium the percentage of people involved in agricultural activities in North America is only in the order of 2% to 3%, virtually all members of a tribe were involved in food generation in the Paleolithic Age.

Then, the Second Wave hit.

## The Second Wave: The Industrial Revolution

For Toffler (1980) the Second Wave is defined and was brought about by industrialization. It began around the middle of the 18$^{th}$ century, but some of its precursors go back further: oil drilling on a Greek island as early as 400 BCE, the existence of money and exchange, the network of trade routes from Europe to Asia, the emergence of urban metropolises in Asia and South America, etc. (p. 38).

Prior to the industrial revolution, most people were essentially self-sufficient; they produced what they consumed. With the Second Wave, peasants lost their autonomy, and production became intended for markets. Toffler (1980) argues that this created "a way of life filled with economic tensions, social conflict, and psychological malaise" (p. 53).

The Second Wave gave rise to the division of labor, which allowed workers and trades people to be better trained, hone their skills in apprenticeships, and gain more experience in what they did. The division of labor culminated with the invention of the assembly line, a production technique which consisted in splitting up an elaborate process into simple repetitive tasks that could easily be performed individually by unskilled labor. Henry Ford is the name most often associated with its introduction in America. The new approach greatly accelerated production and reduced costs.

The greater affordability of goods translated into increased demand, which paved the way for another Second Wave phenomenon and defining characteristic of industrialization: mass production or the large-scale manufacturing of items that are identical. This drove prices further down and gave rise to today's style of consumerism.

### Industrial Technology: The Power to Move Mountains

One of the key elements in the development of industrialization was the multiplication of human strength and power by several folds. For example, to exploit the tar sands in Alberta, Canada, we have created monster trucks having carrying capacities of 360 tons. In comparison, the average vehicle used to deliver landscaping soil to your home carries loads of 1/2 to 3 tons.

The piece of technology responsible for our ability to move mountains is the engine. Just like the water wheel and the windmill early on, the invention served to multiply labor output. It powered a new era, revolutionized transportation, and eventually created a booming automobile industry. The invention of the engine resulted in a massive increase in wealth and productivity.

However, unlike earlier technology that used clean and renewable energies such as the wind and water, the new contraption was fed by fossil fuels produced in ancient times. This signaled a significant shift away from clean sources of power and represented perhaps our first step down the global warming road.

The fossil fuels that we have been tapping into to support our lifestyles for the last century have been a boon but also a scourge in their massive contribution to pollution and global warming. By the early 1970s, the petroleum industry had been so successful in expanding wealth that virtually the entire world economy had become highly dependent upon it.

That is when the first oil crisis hit.

## The Third Wave: The Internet Age

Toffler (1980) argues that up until 1973 industrialization had ruled unchallenged. The Soviet Union and the US were locked in the Cold War, both vying for allies worldwide, seeking to expand their influence and shore up their defenses. Multinational corporations emerged as a third power—often under the protection of their national governments—spreading their tentacles across the planet in their relentless drive for cheap resources and greater profits.

Drunk on cheap and bountiful oil supplies from the Middle East, the developed world saw growing stability and unlimited economic expansion. The new wealth set an upbeat mood. Until 1973, that is, when everything came to a screeching halt.

## *OPEC*

August 8, 1960, argues Toffler, might be symbolic of the final stages of the Second Wave. On that day, in a bid to increase profits, Monroe Rathbone, Chief Executive Officer (CEO) of Exxon Corporation, made the decision to reduce royalties on imported petroleum. The other major players in the industry followed suit within days. Oil producing states—many of which, developing countries—were hit especially hard by the losses. They organized an emergency meeting in Baghdad to address the issue. On September 9, 1960, the Organization of Petroleum Exporting Countries (OPEC) was born.

For many years, it had little success in raising oil prices. However, taking advantage of the outbreak of the Yom Kippur War in 1973, OPEC was successful in cutting production and increasing revenues several folds.

Second Wave production systems were highly concentrated and non-diversified with respect to energy. Transportation was almost exclusively petroleum based—and still is to this day although things are slowly changing. The industrial base of most countries also generally revolved around the same fuel. The rosy outlook for the future was in tatters. The price of gasoline spiked up, sending world economies into inflationary spirals.

In 1980, just as countries were emerging from the first oil crisis, OPEC hiked prices again. The cost of a barrel of oil tripled suddenly to above U.S. $35. The hike in prices resulted in austerity measures being implemented in both developed and developing parts of the world. The new economic conditions brought some countries to the very brink of bankruptcy. In the years that followed, world oil prices dropped back to lower levels. Other cartels do exist. However, none had struck so close to the heart of the industrial economy.

Today, the price of petroleum is up to new heights—over U.S. $130 per barrel, peaking at times past the U.S. $139 mark. Because of the high inflation levels of the last two decades of the 20th century—not to speak of the drop in value of the U.S. dollar—today's prices are actually about 20% to 30% higher than their highest levels after OPEC's second hike in the 1980s, not quadruple what they were.

### *Toffler's De-Massification*

Second Wave industry developed from mechanical inventions and focused on mass production. With the advent of the Third Wave, Toffler saw the pendulum swinging back from the petroleum-based heavy industry towards more appropriate and less oil-intensive technologies, towards a de-massification of production.

He also makes the case that the Third Wave industrial base would operate on a more sustainable basis (energy-wise) and be composed of a mix of *high-stream* industries (large scale and science based) that would be under tighter social controls and friendlier to the environment, and of *low-stream* industries—of smaller, more appropriate, and human scale—that would rely on new and sophisticated technology.

## The Missing Link

Despite Toffler's valiant efforts, we do not have yet the answers to all the questions posed by the 20th century. One of the most important issues in socio-economic change is implementability. No matter how beautiful something looks on paper, it has no use if it is impractical. Neither has it much use or value if the extent to which it works is not enough to solve the problem. In many instances, an insufficient dose of medicine will not cure the disease and may even result in the patient's death.

It should be obvious to all of us that with respect to the environment there is still a piece of the puzzle missing. Despite the scientific advances of the last century, we are not keeping up with problems, let alone solving them.

The earth and the atmosphere are slowly being poisoned by a number of pollutants. Where people do not get sick or die, the contamination continues silently, uninhibited. Environmentalists' efforts are laudable, but they are not enough. Problems are massive and will not get resolved with piecemeal or incremental solutions. This is where this book comes into play. It takes an appropriate-scale and comprehensive approach to problem-solving for the environment.

## The New Millennium

At the beginning of this new millennium, oil remains an overriding concern for modern societies. As we look into the future, scarcity could turn other non-renewable resources into objects of power. This would create international tensions, fuel conflicts, and eat away at our standards of living. Unless we change the way we manage those, we will be looking at a very chaotic future, and the current turmoil may only be a taste of what is to come.

# 3. Energy: The Past and the Future

This chapter provides a general introduction to energy issues and will serve as a basis for the development of the environmental strategy proposed later on. If you already have an advanced knowledge of this topic, feel free to fast forward through this part or jump to the next chapter.

## Ancient Energy

Two defining characteristics at the core of contemporary society are technology and fossil energy. The massive productivity of modern machinery is not only the result of engineering designs but also of energy. The work that machinery produces does not come from human muscles but from fuel.

Most industrial technology is based on petroleum, which has the distinctive characteristic of being a fossil fuel. Oil, like other mineral resources, is free. We really only bear the costs of extracting, refining, and getting it to the gas pump.

We are vastly more successful than Stone Age people not only because of the technology we have developed but also from tapping into vast and inexpensive sources of energy produced in ancient times. When the price of that commodity increases as is happening now, we see how quickly it can affect our standard of living and how much of modern society's success depends on cheap fossil energy: the cost of everything goes up, some countries face food crises, and a lot of wealth simply disappears.

To a large extent, the above are problems resulting from the Second Wave.

## The Carbon Cycle

Oil, gasoline, and other fossil fuels come from biological (plant and animal) matter or *biomass*. Over millennia, vast amounts of vegetation and animal wastes were deposited at the bottom of bodies of water or submerged as a result of one cataclysm or another. Under certain conditions, part of the sediments eventually turned into coal, petroleum, natural gas, or other fossil fuels. In areas of the world where non-permeable layers formed on top of the biomass, even the more volatile elements, such as natural gas, remained trapped.

The question is, what is the vegetation that turned into oil made of? Most of us would say dirt, but we would be wrong. If you remember your biology classes, you know that the bulk of the weight of a tree is composed of only a small fraction of soil. A large part of the total weight of vegetation is water. The other large component of plant matter is carbon, as in carbohydrates or what the body uses to produce energy.

It does not come from the soil but from carbon dioxide gas ($CO_2$), which is a natural component of the atmosphere. It is one molecule of carbon attached to two of oxygen. Breaking these apart takes energy. Recombining them gives energy. With power (light) from the sun, trees break apart the molecules of the gas and steal the carbon for their own growth in a process called *photosynthesis*. That element is the main building material of plants. The oxygen is released into the air and left behind for us to breathe.

When vegetation decays under specific circumstances, it is transformed into hydrocarbons—the main components of petroleum. They are another way in which the carbon originally captured from the air by plants is stored in organic matter.

When you eat and burn the carbohydrates from food in your own body, you recombine carbon and oxygen to reform carbon dioxide, which you breathe out. In the process, you release energy that fuels your body. In the same way, hydrocarbons are burnt in car engines in a process that recombines carbon to oxygen from the air (combustion). The carbon dioxide gas produced is released back into the atmosphere via car exhausts. Of course, as petroleum is not pure hydrocarbon, many pollutants are also created and vented out through the combustion process.

### Global Warming and the Carbon Cycle

Carbon dioxide is a greenhouse gas, or a gas that significantly contributes to global warming problems. It is also one of the main targets for reduction under the Kyoto Accord. It acts as a blanket around the earth and reduces heat radiation into space, keeping the planet warmer. Higher concentrations of greenhouse gases in the atmosphere result in rising global temperatures that threaten to melt the polar ice caps, raise sea levels, flood coastal regions, disrupt global weather patterns, and even trigger a new ice age.

Vegetable oils are also a form of fuel. They burn like petroleum and can be processed to produce biodiesel, which can be used in regular engines. Whether you burn petroleum, coal for electricity generation, wood, biodiesel, or vegetation, you always end up recombining carbon with oxygen from the air to re-form carbon dioxide, which is emitted back into the atmosphere. Growing vegetation reduces global warming. Breathing and burning fuels, on the other hand, increase it.

The carbon dioxide that humans and animals breathe out is not what creates global warming problems. Neither is it the burning of wood. What we eat and the logs we burn are carbon that has recently been taken from the atmosphere by trees or vegetables in their growth process. We are just putting it back in. These activities are *carbon neutral*. One ton of $CO_2$ removed plus one ton put back in equals zero. Based on the same principle, renewable fuels produced from corn or other crops are also carbon neutral.

This is not the case, however, for fossil fuels. They are the major culprits behind the greenhouse effect because they were produced millions of years ago. The mining and pumping out of the large and ancient pools of energy releases the carbon from those deposits into the atmosphere in massive amounts. This results in increasing concentrations of atmospheric carbon dioxide and rising global temperatures due to the blanketing effect. This is worsened by the fact that we are emptying those pools in a relatively very short period of time: decades, perhaps a century or two.

Another significant source of greenhouse gases is the large-scale and permanent deforestation of the planet.  As seen earlier, trees store carbon.  Growing one and cutting it down for fuel is carbon neutral: + one – one equals zero.

However, the global forests are themselves a vast reserve of carbon just like the underground pools of oil.  Although a single tree may live or die, the forests themselves have been around for millions of years. Reducing global forestation without replanting would add new carbon to the atmosphere and increase global warming.  Worse, decreasing the total amount of forested area around the globe has a secondary effect: it reduces the planet's ability to take carbon out of the atmosphere.

The world's oceans play a significant role in carbon absorption. Analyses showed that oceanic waters had absorbed about half of the carbon dioxide emitted as a result of human actions since the beginning of the Industrial Revolution.  This has served to slow down global warming.

However, the absorption has had a dramatic impact on marine life. As carbon dioxide is taken up, it transforms itself into carbonic acid which increases the acidity of water and removes calcium carbonate from oceans.  This compound is needed for the shells of many marine species.  Some of these form the very basis of the oceanic ecosystem. Plankton, for example, is at the bottom of the food chain and is critical to the survival of many species.  Its reduction can have devastating effects throughout the ecosystem.  This also implies that global warming problems could not be solved through the injection of carbon dioxide into oceans (Lean, 2004, August 1).

### Transition Fuels

Because fossil fuels are used so massively in today's society, most of the experts in the field do not believe that a full direct conversion to renewable carbon-neutral energies is going to be possible.  They talk about transition fuels that may be cleaner than what we currently have and could be used in the short and medium term.  Here is a brief look at them.

Coal, which is abundant in many countries, is being investigated as an alternative source of energy for the future. It is already widely used in the industry for steam, heating, and the production of electricity. As fossil fuel, it is not renewable or carbon neutral. Its combustion is a source of mercury pollution around the world and contributes to global warming.

Cleaner burning technology is being developed. The production of liquid fuels (for example, diesel through the Fischer-Tropsch process) and extraction of hydrogen from coal are possible future avenues for exploiting this resource. Although coal may open up possibilities for the transition period, it is not a long-term solution to the energy crisis because, among other things, it is a fossil fuel.

Nuclear energy has seen a renewal in the wake of the oil crises. However, like metals, fissile materials are minerals that are depletable. Furthermore, we still do not have any fully safe options for long-term storage of the radioactive waste produced by the industry. Spent fuel poses a security threat as it can be used to build dirty bombs (standard explosives packed with radioactive material). Although these would not set off a nuclear explosion, they could contaminate a wide area. Multiplying the number of reactors world-wide would also increase the risks of meltdowns.

The use of natural gas has been increasing over the last decades and is expected to continue to do so. Growing interest in this source of energy and more exploration have resulted in an increase of known reserves (Geller, 2003, p. 25). Although natural gas is not a renew-able energy and is depletable, it is many experts' best hope as a bridg-ing fuel. It has good prospects for enabling us to make the transition from petroleum to the renewable energy sources that will power our future.

Natural gas burns much more cleanly than gasoline, diesel, and heating oil. Its combustion also produces less greenhouse gas (i.e., is less carbon intensive) than other fossil fuels and does not release any sulfur dioxide—a toxic agent—or particles. Furthermore, the emis-sions of nitrogen oxide in gas-fired power plants are much lower than those of the new coal technology for electricity generation (Geller,

2003, p.25). Those of carbon dioxide in natural gas power plants are less than half (55% to 65% lower) of those of their coal power equivalents (Geller, 2003, p. 25).

Natural gas is more widely distributed on the planet than petroleum. Therefore, it decreases the world's dependency on Middle East oil and increases global security. A number of mega-projects for the development of new sources of natural gas are under way in many countries around the world. The sheer size of the capital investments already involved may simply preclude turning back the clock on natural gas even though it does contribute to global warming. Next, we will take a closer look at renewable energy options.

## Renewable Energies

Much of the information in this section is drawn from a report by Stelios Pneumaticos, Renewable Energy in Canada. Status Report 2002. Rather than providing a comprehensive view of renewable energies—which is a vast and quickly evolving field of research that could be in itself the subject of an entire book—this chapter reviews some of Canada's efforts in that sector. Selecting a specific country will make more tangible many of the issues involved.

Canada is a world leader in the development of renewable energy sources partly because of an abundance of hydro-electricity. Currently, it produces about 17% of its fuel and power requirements from renewable sources. Following, is a brief overview of the variety of renewable energies in Canada and of some of the issues relating to specific resources.

Hydro-electricity, a relatively clean and renewable energy, accounts for about 11% of the total primary renewable fuel and power production in Canada. However, it is not a new form of energy. Hydro-electricity production saw an expansion as a result of the world oil crises of the 1970s and 1980s. It is expected to continue to grow in the future. Its focus will probably shift to smaller scale technologies, which are more environmentally sound and less disruptive to sport and commercial fishing.

Biomass is the second most important source of renewable energy in

Canada. It accounts for about 6% of the country's primary fuel and power production. It comes in different forms, of which some are not new. Wood has been used directly as combustible since the domestication of fire. To this day, it is still used for heating as well as electricity production.

The technology has evolved, but the principle is still essentially the same: combustion. Slow-burning stoves, wood pellets made from byproducts of the lumber industry, and external air intakes combined with heat exchangers are some of the new technologies designed to make combustion more efficient and take advantage of plentiful and cheap leftovers from the forestry and other industries.

Increasingly, wood derivatives and residues are also used for low-grade heat, steam, and even for the generation of power in the lumber industry itself. Wood is plentiful in many countries. It is a renewable resource as long as the industry is properly managed. It is carbon neutral but not really a clean energy.

Having a few houses in the countryside with wood stoves is one thing, but an entire city heated by firewood is something else altogether. It would vastly add to the existing pollution problems that many urban areas already face. Smog problems would be significantly aggravated. Montreal, Canada, for example, sees many days of *winter haze* on account of the increasing use of wood stoves. Some small towns in Ontario have been experiencing air pollution levels similar to Toronto's for the same reason.

New advanced catalytic combustion stoves are believed to reduce particle emissions, smoke, and other pollutants by about 80% compared to earlier models. Whether or not this new technology will make feasible their use on a large scale remains to be seen. Included in the biomass category are also solid wastes from both municipal (garbage) and industrial sources. These can also be burnt to generate steam, heat, and electricity.

Biomass is also bio-gas. Methane—an alternative to natural gas—is increasingly produced from a variety of organic byproducts and leftovers from a number of sources and industries. The gas can be collected from landfill sites, generated from municipal sewage, or produced through anaerobic fermentation of agricultural products and byproducts.

Gasohol, E85, and ethanol blended diesel are other forms of biomass energy. Gasohol is a mix of about 90% gasoline and 10% ethanol, or regular drinking alcohol. It is cleaner and deemed to be a superior fuel for winter driving. Gasoline is a fossil fuel. However, ethanol can be produced from several types of crops, for example, wheat, barley, and corn.

Unfortunately, these are the same grains that can be made into flour for baking, and, as we are seeing now, the production of biofuels can contribute to a rise in the price of food. Ethanol blended diesel contains 10% ethanol. It is cleaner burning than the fossil fuel alone. Both gasohol and ethanol blended diesel can be used directly in regular engines.

E85 is 85% ethanol with only 15% gasoline. Its use requires a gasoline engine that has been modified. Some U.S. automakers have already begun to make hybrid engines that can burn all, gasoline, gasohol, and E85. The production of liquid fuels from agricultural residues and non-food crops grown specifically for energy (e.g. switchgrass and trees such as willow and poplar) has attracted growing interest. Some of the issues related to biofuels are the initial capital costs (for example, equipment), the development of a common distribution system (i.e. gas stations), and the competition with food crops for land.

According to the California Energy Commission web site (http://www.consumerenergycenter.org/, April 13, 2004), Brazil now has some four million cars running on ethanol. This use of renewable fuels is the result of a government program aimed at decreasing the country's dependency on fossil fuels and producing energy domestically from sugar cane, a local and plentiful crop. The site also reports that the quadricycle, one of Henry Ford's first automobiles, was powered by pure ethanol. His Model Ts could also be run on this fuel, having a carburetor that could easily be adjusted for ethanol.

American transportation could have been fueled—and could still be today—by energy produced by American farmers for the benefit of the entire country. That would have prevented their current dependency on the Middle East. Brazil—a country much less well off than the US—has done it, proving that renewable energy strategies are realistic, feasible, and can be afforded by many countries. As seen above, some of the technologies involved have existed for a long time. Had the world followed the example of Brazil when it went through its own transition, global warming would not be the issue it is today.

Biodiesel is an alternative to regular diesel. It can be made from vegetable oil. Many crops are suitable for its production, for example, soy, sunflower, canola, hemp, etc. It can be used in pure form or blended in different proportions with regular diesel. This alternative can often be fed directly into existing diesel engines with little or no modifications. Unlike fossil fuels, the vegetable oil alternative is carbon neutral and does not contribute to global warming. It also burns more cleanly. Biodiesel combustion produces fewer gas and particulate emissions and lower levels of carcinogens (Pinderhughes, 2004, p. 176).

Wind provides one of the fastest growing and most promising sources of clean and renewable energy worldwide. Turbines are erected in fields to capture the energy from the wind and transform it into electricity, often feeding it directly into the existing electrical grid. However, wind is only an intermittent source of energy. When it does not blow, no electricity is produced. This somewhat limits its usability in that electricity has to be stored or supplemented. Depending on how storage technology develops, this may become a very viable and widespread source of alternative energy.

There are also a number of smaller players in the renewable energy field. For example, tidal power—energy captured from waves and the rise and fall of tides—is clean and renewable but is obviously limited to coastal areas. Geothermal power, or energy extracted from the earth's crust, has fair prospects in a number of locations around the world. It is essentially the same energy as that responsible for hot springs.

Earth-energy systems (EES) are a slightly different kind of geothermal energy. Heat pumps extract low-grade energy from large earth subsurfaces that are warmed by the sun or underground water sources. Conversely, if cooling is needed, heat is extracted from a room and sunk below the ground. EES systems are capital-intensive technologies.

Solar power is probably the most well-known source of renewable energy. It attracted attention when people first began considering alternatives to fossil fuels following the oil price hikes of the 1970s and 1980s. It is currently used for both space and water heating in both domestic and industrial sectors worldwide. The US, Japan, Australia, and Israel are considered leaders in this market.

There is a variety of systems for capturing energy from the sun. They fall into two categories: passive and active. Passive solar systems primarily capture heat from the sun through windows or sun-absorbing matter for direct space or water heating. Historically, windows were energy inefficient. They provided lower insulation than walls. However, new technology that offers better heat loss prevention and lower emissivity has improved windows to the point where advanced models now provide a positive energy supply.

Solar photovoltaic energy (PV) is an active type of power. Semiconductor technology is used to transform sunlight directly into electricity. At this point, PV is still too pricy to replace conventional sources of energy. However, it has uses in remote areas in which there is no existing electrical network and in stand-alone applications. Examples of these are remote residential locations where grid extension would be expensive, road signage, coastguard systems, and remote monitoring. The main challenge for PV as an energy for the future is cost reduction. New technological developments and production automation are avenues through which more competitive prices may be achieved.

Solarwalla—usually perforated metal paneling mounted against buildings on vertical south-facing walls—is technology designed to absorb solar energy for air and hot water heating as well as for cooling in industrial plants. It is gaining grounds in renewable energy markets because of its cost-effectiveness and the fact that existing structures can easily be retrofitted with the technology. It can provide up to one third of a building's heating and air make-up needs. Solarwalla is also being investigated internationally for use in other applications such as commercial and agricultural drying.

There are many other solar technologies—both passive and active —in use and in development for various applications: electricity generation, residential (including pools) and commercial water heating, and air make-up.

### Renewable Energy Issues

Renewable energies have significant advantages over fossil fuels in that they are generally cleaner, unlimited, and do not contribute to global warming problems. However, many are not as concentrated or portable as petroleum. These two qualities are important for their potential replacement of fossil fuels in the transportation industry. Electrical power may be suitable for commuting to work, but storage capacity is still low at this point in time, making it impractical for heavy-duty and long-range transportation. There may also be temperature issues in cold climates.

Technological development is not the most important obstacle to the widespread and large-scale use of alternative energies, to their replacing fossil fuels as power base for economies around the world. Up until very recently, ups and downs in oil prices were the problem. We saw bursts of R&D, government subsidies, and investment in alternative fuel technologies in the wake of the oil crises of the 1970s and early 1980s. Many of these were subsequently lost when the cost of petroleum dropped.

Many types of renewable energy are only profitable when the price of oil remains high. As such, unpredictable petroleum prices were a stumbling block for the industry in the last two decades. This is now changing as most experts in the field believe that the cost of oil will stay high. A more recent problem is the rise in the price of food from the use of edible crops and good agricultural land for the production of biofuels.

Another concern is that converting to new energies often requires a substantial investment in infrastructure. For example, gas stations need to be modified to accommodate new fuels or a new distribution infrastructure might have to be built parallel to the existing one.

As I put the final touches on this book, this picture is quickly changing. By the time you read this, oil prices may already be high enough —although they might not remain there—to foster the development of a strong renewable energy sector.

## *Methane Hydrates: Energy of the Future?*

Methane hydrates have been heralded by some as the possible source of energy for the future. They are gas molecules trapped in ice on the ocean's floor. Research is still very preliminary at this point in time. Reserves estimates range from a few hundred to a few thousand years. Methane is considered a relatively clean burning gas.

There are a number of issues relating to the exploitation of hydrates. Firstly, they are not carbon neutral and pose the same problem as other fossil fuels in that respect. Secondly, there are the logistics of mining a resource several hundred meters under the surface of the ocean. Thirdly, methane is a greenhouse gas 10 times more powerful than carbon dioxide in its contribution to global warming. How much of it would escape into the atmosphere as part of the extraction process? Furthermore, it is believed that mining activities may destabilize the ocean's floor and cause landslides that may disturb hydrate deposits and result in the release of vast amounts of methane into the atmosphere. Many questions remain to be answered at this point.

Research is currently looking at carbon-neutral ways to exploit the resource. The methane removed would be replaced with $CO_2$ hydrates. Visit the following web site for more details: http://www.eee.columbia.edu/research-projects/sustainable_energy/Hydrates/index.html. Whether the new exploitation methods prove feasible remains to be seen. It would certainly much enhance the value of this resource.

Although methane hydrates could serve as transition fuel, they are not the solution to long-term energy issues. In terms of global warming, we certainly cannot afford to add another 3,000 or 4,000 years' worth of greenhouse gases to the atmosphere. If carbon-neutral methods are developed, our energy reserves would only be extended temporarily, and we are not even there yet.

The real long-term solution to energy problems is to switch to renewable sources. Because of global warming issues, the discovery of a gigantic new reserve of oil, gas, or other fossil energy would not solve our problem. We have to come to terms with the fact that fossil fuels will likely need to be phased out regardless of reserves. Carbon-neutral means of exploitation can help but will not solve problems in the long term.

We also need to change the attitude that it is our right to wipe out one resource after another in order to fulfill our selfish needs and prop up our standards of living... leaving nothing to generations that will inhabit the planet only a few hundred years from now.

### The New Global Warming Equation

Even with a successful Kyoto Accord, we would still continue adding huge amounts of greenhouse gases to the atmosphere. It would not solve the problem, only alleviate it. Global warming is ultimately a function of two factors: how much greenhouse gas is added to the atmosphere and how much is removed from it. We increase global warming not only by burning fossil fuels but also by deforesting the planet. We reduce it by growing vegetation.

Biomass has been stored over millennia not only as fossil fuels but also as live plant matter. As discussed earlier, when we harvest trees for one industry or another and reforest afterwards, we do not add greenhouse gases to the atmosphere. Over the medium and long terms, carbon is re-stored into the biomass of the new trees. However, when we clear cut a forest without replanting, we reduce the total amount of plant matter on the planet. When this occurs, atmospheric greenhouse gas levels are increased exactly as if fossil fuels had been burnt because the carbon added to the atmosphere is never re-stored into new trees and forests.

If the total live biomass on the planet increases, it could compensate for some of the fossil fuel combustion we are doing. The problem is that it is actually decreasing, not only adding to global warming itself but also reducing our capacity to absorb carbon from the atmosphere or compensate for fossil fuel use. The more trees there are, the faster carbon can be absorbed back into biomass. This capacity has been decreasing in the last few decades as deforestation has occurred in many countries, the clear cutting of the Amazonian forest being only one of many examples.

With the discovery of methane hydrates, we now have a third component to the global warming equation: the indirect release of potent fossil greenhouse gases. As the earth warms up, we can expect that water will also see a rise in temperature. As this occurs, the massive beds of methane hydrates at the bottom of oceans could begin to thaw out and release into the atmosphere large amounts of a greenhouse gas which is 10 times more potent than carbon dioxide. This will accelerate the greenhouse effect, which will result in the release of more seabed methane and the speeding up of global warming. This is a vicious circle that may make things happen much faster than anticipated.

There are also fears that global warming will result in the thawing of the permafrost in northern regions, releasing the millions of tons of carbon dioxide trapped in it, further increasing greenhouse gases in the atmosphere. So, there might just be a fourth and significant component to the global warming equation. How many other factors are yet to be discovered?

We should get used to the idea of increasingly violent and temperamental weather patterns and their happening potentially much sooner than we think. Sea levels will likely rise much faster than predicted. The flooding of coastal areas would also occur much sooner than anticipated. Rebuilding New Orleans may turn out to be a major mistake. Despite all the new diking and damming, there are very good chances that the city will be under water again. In view of even only the last two components of the global warming equation, we may want to do everything we can to speed up our shift to renewable energies.

# 4. The Resource Conservation Failure

## The Rise of Consumerism

Technology grew by leaps and bounds through the 20th century. A large number of inventions have enabled us to produce all kinds of gadgets to make our lives easier, to make our leisure time more interesting, more entertaining.

But, has that technology run out of control? Producing more and more means that we are using up more and more non-renewable resources. And, we are now over six billion consumers on the planet. According to Malcolm McIntosh (2000), a writer, broadcaster, and lecturer on corporate responsibility and sustainability, "Since the mid-20th century the world has consumed more resources than in all previous human history" (p. 47).

Minerals are limited in supply and do not belong to us alone but also to future generations. In but a few decades, we have used up several times our share. What is our plan for the next half century? Double that?

At the end of the last century, we were using up resources a lot faster than in the 1950s. We might equal the amount of resources used between 1950 and 2000 in only the first 20 years of the 21st century, or less for that matter. Worse, the rate of use is only accelerating.

China has a population of about 1.3 billion and has been exporting goods for a long time. However, not until recently has it seen enough income growth to support a significant amount of consumption. With its recent wave of trade and economic liberalization, things are changing fast as individual income grows.

India is not far behind in terms of population, and its economy is growing equally as fast. Both are entering an era of consumerism. Together they represent about one third of the entire world population, currently estimated at 6.5 billion. Can you imagine the amount of resource depletion that will occur when these two countries alone continue to grow? In the next 50 years, we are likely to wipe out three to five times the total amount of resources consumed in the second half of the last century.

## The Corporate Solution

Corporations never disappear, only resources do. As these become scarce, businesses will continue to sell them to us or move to something else. In fact, the oil experience has shown that shortages often result in higher prices and greater profits for them.

What they do is use of the cheapest, most economical resources first. For example, the cheapest oil is pumped out first and used up. Then, the next cheapest source is used. Once it is exhausted, they move on again, so on and so forth.

As the price of resources goes up, substitutes that were more expensive or not as suitable become profitable and can be exploited. For example, natural gas, which is generally more difficult to handle than petroleum, would replace oil as it becomes scarce. When natural gas reserves suffer the same fate as petroleum, the market would again look for the next less suitable or pricier alternative. So on and so forth.

## Problems With the Business Resource Model

The beauty of the above model is that corporations will still make profits when the price of a liter of gasoline reaches $10.00 ($40 a gallon). In the 1970s the giant oil corporations were criticized for price gouging. As I edit this book, the headlines on CNN/Money read, "Big oil CEOs under fire in Congress. Lawmakers spar with execs from Exxon, Chevron over high prices, record profits, consumer pain" (Isidore, 2005, November 9). If you follow stock market news, you will notice that when the cost of a barrel of oil goes up, the price of petroleum industry stocks increases, signifying an expectation of greater profits.

Corporations will make profits selling us gasoline at $2 a liter. They will also do so when its cost reaches $20 a liter. We, however, will be the ones picking up the tab for this. Resource depletion is not good for us. We will be left with increasingly higher prices and lower standards of living. The business model will lead us to use up resources one after the other as if there were no tomorrow and with total disregard for future generations.

## Planning for the Future

Successful planning anticipates problems and fixes them before they arise. We need to plan ahead and conserve the mineral supplies we have to prevent a catastrophe from happening. Once non-renewable resources are gone, they are gone. Depletion is irreversible and will leave future generations with high resource costs, dysfunctional economic structures, and much lower standards of living. The stakes are high.

Actual estimates of reserves of different minerals vary not only from year to year but also according to technological developments, politics, and geopolitics. Recoverability and prices are also variables that make it difficult to determine accurately how long resources will last. The issue will be discussed in more details later. Suffice it to say that the only absolute for non-renewable resources is that estimates are mostly only in the dozens and hundreds of years, not the thousands and millions of years that they will be needed for. In the long term, there is only one trend: reserves will decrease and prices will rise.

Known oil reserves were extended as a result of a number of scientific discoveries, processing innovations, new exploration efforts, etc. Petroleum prices may have dropped or their increase slowed down on a short-term basis as a result. However, most of these factors do not have a significant impact over an extended period of time although they can lower costs temporarily.

The consensus among economists and experts with respect to oil is that its price will only continue to go up in the long term. Liberal estimates are that reserves will peak in 10 to 20 years. Some market analysts even argue that they reached their highest levels around 2004-2005.

## The Case of Energy

Energy looks like a poster child for the business resource model because of its substitutability. Although oil itself is not renewable and will run out eventually, it is highly substitutable. That is, when it and other fossil fuels are gone or get to be too expensive, we will shift to energies that are plentiful, unlimited, and fully renewable. That model only works because many of the long-term alternatives to petroleum and other fossil fuels are good and relatively inexpensive substitutes.

A number of these are already in use. You do not see them yet in the transportation industry mostly because of price, infrastructural, and technological issues. Hydro-electricity is inexpensive and plentiful in many countries but not often used in transportation because of driving range limitations at the present state of technology. Energy from wind and biomass is also essentially unlimited and will one day contribute to replacing fossil fuels.

The world can actually get by on hydro, wind, biomass, and solar energy alone. It could do that not only forever but also at a relatively low cost.

The business model does not work with other mineral resources. Its fundamental aspects are true: use the cheapest source first, and move on to the next cheapest one after. Science has also led to greater efficiencies. But...

## The Case of Metals

One problem with the business resource model as it pertains to non-renewable resources is that energy is one of the few fields where the theory works. Would all other mineral resources have the same characteristics as energy, everybody would be jumping for joy. There would be no resource conservation issue. However, that is not the case.

## *The Substitution Argument*

Steel, aluminum, and copper can be substituted for each other in many applications. They are mainstays of the modern world, being used everywhere in buildings, electrical infrastructure, and a variety of consumer goods. Although there is the possibility of substitution, these three metals are all not renewable. They are all being used up at the same time and would generally see their costs steadily increase as time goes by. If their depletion rates and costs go up in parallel to each other, then they are not true substitutes. That is, one could not replace the other in case of depletion.

For example, if 50 years from now reserves of iron have been exhausted, you would not be able to switch over to a plentiful supply of aluminum because it would also have been used at the same rate and be in shorter supply or near depletion by that time. We might even have already been considering switching from aluminum to steel as our supplies of the lighter metal ran low. If a metal is not renewable and if it is being depleted at a similar rate, it is not a true substitute. It would not solve any shortage problem.

Furthermore, an excessively priced alternative—as would be the case if its supplies were being depleted—would be useless. A true substitute needs to be both plentiful and reasonably priced at the time of substitution, i.e. not now but when the first resource is becoming exhausted.

## *The Issue of Massive Use*

There are no true substitutes for most common metals because they are all mainstays of the modern world and massively used. As such, alternatives would have to be available in enormous quantities at the time of substitution. For a while, we might be able to switch to them. However, that would only be a temporary measure. The substitute would then be used up twice as much and at twice its earlier rate. It would be quickly depleted, giving us but a very short reprieve, if any. There are no real substitutes for many of the basic materials on which society's infrastructure is built. Their massive use underscores our fundamental dependency on them.

Let's take a real life example to illustrate the above. Suppose that three people took a trek in the desert. They each brought along a limited resource, one bottle of water.

Wondering whether there would be enough water for the trip, the leader takes a look at the other two hikers and sees their bottles. He concludes that, when his supply runs dry, he can just help himself to the other two bottles which are quite full.

Of course, when the first one is empty, there will not be much left in the others. When two people drink from the second bottle, they will exhaust the little that remains in it twice as fast. It will not last very long. What is left in the third bottle will be exhausted even more quickly when three people depend on it.

In the real world, there will not be a substitution or jumping from one resource to another and another ad infinitum into the future. There will be a gradual increase in prices until that process starts to accelerate. By then, it is going to be too late. Panicked substitution will barely mitigate the problem and only last for a short time before reality comes crashing down on us.

## The Suitability Issue

Although metals may theoretically replace each other in many applications, they are not necessarily good alternatives. For example, gold, steel, and lead could probably all be used in electrical wiring but would be poor substitutes for a number of reasons. Gold would be excessively expensive, steel would lack flexibility, and lead is toxic.

Most metals would actually be very poor substitutes for each other because of economic and suitability issues. Their use as such would create a dysfunctional society and be the result of the actions of very desperate people. Furthermore, this would probably occur at a point when society, itself, has already reached a state of economic crisis. The question is, do we want to wait until then to start conserving resources, until it is too late? Socio-economic dysfunctionality is likely to increase as resources are being exhausted.

To a large extent, substitution is a myth when talking about non-renewable resources. For us to live in a fantasy land with illusions of unlimited resources and substitution is dangerous. The reality is that, rather than jumping from one mineral to another as they are being exhausted, the world will run out of most metals almost concurrently. When that begins to happen, panicked and dysfunctional substitution will make little difference.

## The Issue of Resource Ownership

Another problem with the business model is the issue of price and ownership of non-renewable resources. These do not belong to us alone but to all generations the planet will see over its lifespan. The business model blindly skips over that part. It assumes that all these resources are ours to waste at will and with reckless disregard for anyone else coming after us. We ought to use all non-renewable resources as frugally as possible.

## The Scientific Breakthrough Argument

Another major problem with the business resource model is the scientific breakthrough argument. One of the most powerful forces behind the fantasy world of unlimited non-renewable resources is the belief that science will solve all our problems. It will undeniably enable us to find new supplies, will help in developing ways to get more of existing supplies, and help keep costs down. But, it will fall short of the illusion it has created.

The business resource model is based on the assumption that future scientific discoveries will be made and that we can waste non-renewable resources in anticipation of that. By doing so, we seriously mortgage our children's future. Planning should be based on reality, not on fantasies and mere possibilities. The current state of science is that minerals (except energy) are not renewable and do not generally have true substitutes.

## The Easy Science Issue

Science behaves to some extent like an exhaustible resource. For example, in exploiting minerals, the most plentiful and easily accessible deposits are usually wiped out first. In several countries, including the US, many of the mines closest to population centers have already been depleted. As a result, resources are increasingly found further and further away and are more and more costly to process and bring to markets.

Science follows a similar pattern. Centuries ago, few things in the physical world were understood. Discoveries that appear to be insignificant today—the invention of the wheel, the mastering of fire —were major breakthroughs that changed the dynamics of entire societies. Today, we are much further ahead. The easier breakthroughs have already been made.

Scientific fields still in their infancy—the computer and information technologies, genetics, medicine, biotechnology—will continue to see new and exciting developments. In the physical resource field, where knowledge is at a more mature stage, we should expect the breakthroughs to generally come less easily and less frequently as time goes by.

## The Five-Billion-Year Question

The earth was formed approximately five billion years ago. Its remaining life expectancy is about another five billion years. Ultimately, non-renewable resources should have to be managed in such a way as to last that long as they also belong to the generations that will live at that time.

In the last 50 years, we have used up as much of the earth's resources as have all the generations before that. In the same period of time, we have depleted maybe 25% of the known oil reserves. Experts estimate that in about 10 to 20 years these will have peaked and will begin to decline. In total, the bulk of world oil reserves will have lasted maybe 300 to 500 years. If the earth's lifespan had been 24 hours, our oil reserves would have lasted less than a second!

The petroleum experience has shown us that a long time before reserves decline, commodities become the object of politics and see significant increases in price. Other mineral resources will likely follow a similar pattern.

There are two very important distinctions between metals and fossil energy. One is that they do not have true substitutes as oil does. As such, you can expect much steeper price increases. The other is that if we wipe them out, there will not be a second chance because of their lack of substitutability.

## Manganese Nodules: Panacea or Temptation?

First discovered in 1803, manganese nodules are potato-size nuggets of rocky material containing manganese, iron, and a number of base metals. They are found in many sites around the world, thousands of meters below the ocean's surface. They lie in large seabed deposits and in significant quantities.

They are seen as a potential source of ore for the future as reserves of surface metals become depleted. Manganese nodules could be a renewable resource as they are believed to be formed by bacteria depositing minerals from sea water onto their surface. They grow very slowly, at a rate of about 2 mm per 1,000,000 years. Their renewal speed is believed to depend on the amount of surface available to receive mineral particles. Mining them will reduce the total area for depositing and result in slower growth rates.

There are many issues with respect to their exploitation. Firstly, there are environmental concerns. There are also questions about our ability to extract minerals two to five kilometers below the ocean's surface. Their exploitation may turn out to be uneconomical or simply unfeasible. As manganese nodules only contain certain metals, they would not solve all our problems. Their excessively slow growth may mean that the new resource is to some extent depletable. Lastly, this may be our last frontier in terms of mineral reserves. We may want to preserve it for future generations and formally set it aside until we have reached a certain point in the future.

It is difficult to estimate how long existing surface resources will last at our current rate of use. Forecasts are from a few decades to a few hundred years. If we were to preserve enough resources for only 1% of the remaining lifespan of the earth, we would have to stretch what we have another 50 million years!

We cannot even cope at this point with managing or preserving resources that are renewable. World species are dwindling and disappearing. We are totally impotent at preventing deforestation, be it in Nepal, India, or the Amazon basin. The cod fishery in Eastern Canada has all but been wiped out. Seabed resources are probably the only thing future generations will have left after we are done. The last thing we want to do at this point is to move into that last frontier. The solution to our problem does not consist in wiping out one resource after another.  It lies in bringing ourselves under control.

## Managing Resources for the Present and the Future

The earth has been around for five billion years, humans have been around for less than 10 million years, civilization has been around for under 10,000 years.  The planet has another five billion years to go.  In less than 500 years, we will have practically wiped out petroleum resources on the planet.  Other mineral resources have already seen their prices rise and are but a few hundred years behind, decades in some cases.  If humankind totally depletes the earth's resources in another 1,000 years, we will have wiped out the necessities for intelligent life on the planet in less than 0.00001% (one hundred thousandth of one percent) of its entire lifespan, in the blink of an eye.

So, what do we do?

# 5. The Silent Poisoning of the Earth

Over the last few decades, we addressed some of the most obvious environmental problems. However, our efforts have often but touched the surface, dealing mostly with only the worst crises once lives are at stake. Many less poisonous elements—but extremely damaging because of their pervasiveness—are slowly but surely accumulating in the environment.

This section provides a brief overview of some environmental contaminants. Its intent is not to cover the field in a comprehensive manner but rather to give some perspective to the pollution debate, show the extent of contamination, and support the case I make about the slow poisoning of the earth. For those who may want more details, there are many works published on the subject, among others, Nadakavukaren's *Our Global Environment: A Health Perspective* (2000). Those already familiar with the subject should feel free to read selectively.

## Historical Perspective on Contaminants

We have known for centuries that lead, mercury, and asbestos were health hazards. They are not new enemies but old foes. For example, author and lecturer in environmental health, Anne Nadakavukaren (2000) writes,

> Hippocrates described the symptoms of lead poisoning
> as early as 370 B.C.; mercury fumes in Roman mines in
> Spain made work there the equivalent of a death
> sentence to the unfortunate slaves receiving such an
> assignment. (p. 225)

The 20th century saw the development of even more toxic substances. Scientific progress, rapid economic growth, and mass production spurred on the phenomenon. They introduced to the environment an entirely new array of compounds. Many of these are now part and parcel of our lives, found everywhere from the Arctic to the Antarctic, to the tissues of human adults and the unborn. They were and still are spewed out of smokestacks or flushed down our rivers on a daily basis. Others are accumulating at waste disposal sites.

### Polychlorinated Biphenyls (PCBs)

First manufactured in 1929, PCBs are extremely stable in the environment and better known for their use in electrical transformers and capacitors. They found their way into the environment through electrical equipment catching fire, the burning of certain types of wastes, and illegal dumping into waterways by unscrupulous corporations trying to avoid disposal costs.

PCBs are toxic to several species at low concentrations and result in a variety of birth and health problems, including liver disease and cancer. Recent research points to their causing endocrine problems in humans and significant damage to developing embryos and fetuses.

PCBs have the ability to bioaccumulate, i.e. to concentrate up the food chain, from preys to predators and humans. According to Nadakavukaren (2000), in early research "virtually every tissue sample tested, from fish to birds to polar bears to animals living in deep sea trenches, contained detectable levels of PCBs" (p. 232). A US study of breast milk found that 99% of their post-1976 samples tested positive for polychlorinated biphenyls (Nadakavukaren, 2000, p. 232, 234).

PCB production and use in open systems were banned in the US in 1976. However, the toxic compound is still legal in closed operations. As such, they still pose a threat today. How much is left out there in warehouses and equipment in the custody of corporations?

How much will eventually be leaked into the environment or get dumped illegally?

## Dioxins

Dioxins are a large group of chemicals related to PCBs. One of the main sources of contamination is the incineration of medical wastes. In the process, toxins become airborne and are sometimes carried over long distances. Eventually, they fall back down to earth and contaminate soil, plants, and bodies of water.

Dioxins are believed to cause a number of health problems, including chloracne, immune system interference, and fetal toxicity. Although cancer is also on the list, hormonal dysfunction is now increasingly believed to be the toxins' leading long-term concern.

Regulations have resulted in significantly lower levels of dioxins in the environment. They are often present in fatty tissues and foods like eggs, dairy products, and fish. Their consumption has resulted in most people having accumulated in themselves minimal levels of the chemical over time (Nadakavukaren, 2000, p. 240).

## Asbestos

Asbestos' reputation as killer is well established. Nadakavukaren (2000) reports that 30% to 40% of the current and retired asbestos workers who have been exposed to large amounts of the mineral are expected to die of cancer (p. 243). Many others will suffer from asbestosis, a crippling lung disease.

Asbestos is a fibrous mineral found around the world. It has seen a variety of uses in the industry throughout the ages. It was used in the Stone Age by potters for clay reinforcement. During the Roman era, it was woven into fabric, its fire-resistant properties conferring to clothing an element of magic (Nadakavukaren, 2000, p. 243).

According to the US Environmental Protection Agency (EPA), about 700,000 buildings (residential as well as commercial) in the US contain some of it in friable form. The EPA further estimates that over 6,000,000 children and teachers may be exposed to fibers everyday in schools (Nadakavukaren, 2000, pp. 243, 246).

Given the mineral's record, you would expect that its use would have been completely banned, but its production has only shifted from developed to developing countries. India, Pakistan, Korea, and Indonesia are important producers. Brazil's consumption of asbestos was growing at a rate of about 7% annually in 2000 (Nadakavukaren, 2000, p. 249).

### Lead, Mercury, Vinyl Chloride, Fire Retardants, and Jet Fuel

Both mercury and lead have been known for a long time as health and environmental hazards. Romans once lined their wine casks, cooking ware, and aqueducts with the latter. Lead poisoning can lead to mental retardation and death. Its main use today is in car batteries. Nadakavukaren (2000) reports that more than three million tons of it are mined every year and that "not surprisingly, lead is now found throughout the environment—in soils, water, air, and food" (p. 250).

Mercury, the quicksilver of ancient times, has been known and used for more than 2,500 years. It can damage the liver and kidneys and is believed to be responsible for a number of nervous system ailments. Mercury bioaccumulates and is found throughout the environment especially because of its ability to evaporate. It is present in many fish species and continues to be added to the environment from, among other things, the combustion of coal, the incineration of medical wastes, and the smelting of some ores.

The *mercury machine*—all the activities surrounding its production and release into the environment—will not stop on a dime. Even with a concerted effort to address the problem, we can expect the contamination to continue for a long time.

Another significant environmental concern is vinyl chloride. It is a known carcinogen that is released into the air and ground water as the millions of tons of PVC plastics (polyvinyl chloride) we produce every year break down in the environment (Markowitz, 2002, pp. 9-10).

Brominated fire retardants are often sprayed on plastics to reduce their flammability. They are thyroid toxins which bioaccumulate and persist in the environment. A study by the Environmental Working Group (EWG) in the US found that these chemicals were present in surprisingly high concentrations in all their samples of American women breast milk (Lunder and Sharp, 2003, September 23). More information on the study can be found at the EWG web site (www.ewg.org/).

In 2004, perchlorate—a rocket fuel ingredient linked to thyroid damage—was found to be present in cow milk in California and in the drinking water of almost half the states in the US (*Rocket fuel*, 2004, June 22).

As seen in the few examples above, the earth is rapidly being poisoned. We already live in a chemical soup, one that not only pervades the environment but also permeates our bodies through and through. We are leaving our children and grandchildren a planet that is highly contaminated, and many toxic compounds are expected to continue to accumulate in the environment.

## The Conspiracy of Silence

The responsibility for the current environmental crisis does not lie solely with some corporations. It extends beyond them. We keep silent while the earth is slowly being poisoned.

We have had some success with regulations and incentives, but they do cost money if not directly, then indirectly. Governments are not interested in making a costly commitment to the environment if it means that they will be voted out in the next election. We have to play our part in this.

In the last few decades of the 20th century, there was a lack of funding commitment. As a result, little got done, and we failed to bring environmental issues to a head.

## The Fourth Wave in the Making

There is a missing link between the green society that we need to achieve and the rallying cry of environmentalists. There is a reason why their efforts have remained largely fruitless, why we are destroying the environment for our children instead of preserving it, why we are decimating not only our share of resources but also those of future generations.

Part of our failure is due to the lack of real commitment—money

—and the worship of current lifestyles. However, the most important reason for our failure is that we lack an economically-viable large-scale strategy that could turn things around for the environment.

The Third Wave has run its course. It's time for a new tide of change. We need to create a new economic environment that will bridge the gap between theory and practice, turn wishful environmental thinking into action, and quickly reshape the current economic system into a mean lean green machine.

# 6. A Comprehensive Environmental Strategy

Commitment to the environment has picked up but still falls a long way short of what we need. We have to shift gear, or we will be leaving behind a devastated world to future generations. We need fundamental environmental change.

## The First Principle

Many environmentalists have sounded the alarm bell with respect to pollution and environmental degradation, but not as many have campaigned or called for action on the conservation of non-renewable resources for future generations. Yet, these issues are both part of the same agenda. They are intimately connected, one being a significant part of the solution to the other.

One of today's generally accepted principles in science is that *nothing is created nor destroyed*. Antoine Lavoisier (1743-1794) and Mikhail Lomonosov (1711-1765) are generally credited with its development and formulation. The *law of mass/matter conservation*, as it is known, pertains to the fact that things in the physical world are not destroyed when they burn or decompose. They simply transform themselves into something else.

As a kid, you saw a log of wood burn and assumed that the combustion process destroyed it almost completely, that most of its matter simply disappeared. In fact, it was only transformed. Some of it was released into the atmosphere as water vapor, gas, or smoke. Some of it burned down to ashes, and the energy that was captured from the sun through photosynthesis was radiated back into the environment.

According to the Lomonosov-Lavoisier law, the tons of ore that we mine every year are not magically destroyed after we are finished with them. They end up in the environment. Every ounce of every ton of the ore we mine is transformed and dispersed throughout the environment.

Some of what we extract from the ground becomes refuse at mining sites. Some is emitted into the air or discharged into toxic lagoons during processing. Jared Diamond, an American evolutionary biologist and UCLA professor, describes the hardrock (metals) mining industry as "currently the leading toxic polluter in the U.S., responsible for nearly half of reported industrial pollution" (Diamond, 2005, p. 452).

The rest of what is extracted in the mining process is made into goods that eventually end up in landfill sites. But the story does not end there. As Annegrete Bruvoll (Head of Research, Unit for Energy and Environmental Economics, Statistics Norway) reports, "End treatment, i.e. waste disposal and incineration, results in emissions of toxic pollutants and greenhouse gases, and seepage from waste disposal sites pollutes ground water and watercourses" (Bruvoll, 1998, p. 16).

The total amount of minerals we extract every year is exactly equal to the amount of wastes we generate (refuse, fluid discharges, garbage, gases and solids from incineration, etc.). One ton of extracted minerals today means one ton of waste added to the environment tomorrow. We dig minerals annually by millions of tons. Therefore, by the same millions of tons, we create wastes and pollutants every year!

Reusing and recycling are the closest thing we have come to in terms of a solution to environmental problems. They are so far the most we have been able to achieve. Unfortunately, our best efforts only minimally delay environmental degradation. Most things can only be reused or recycled so many times. They all eventually end up in our landfills and the environment.

In the long term, our two most prominent achievements with

respect to the environment, reusing and recycling, will only give us a bit more time as everything will sooner or later end up in the environment. Although their impact is very significant in the short term, they cannot bring change on the scale needed and are not the permanent solution we are looking for.

The more mineral resources we extract, the more pollution and environmental contamination we create. Conservation would achieve two goals at the same time and be part of a long-term approach to the problems at hand. Reducing the extraction of resources would preserve them for future generations, and every ton of minerals not mined would be one ton of wastes and pollution not created.

The first principle of a comprehensive environmental strategy is the recognition that a significant part of environmental contamination and degradation is directly related to the amount of non-renewable resources we dig out of the ground. Decreasing our consumption of these resources is key to both conserving them for future generations and reducing environmental degradation.

## The Second Principle

The second principle of a comprehensive environmental strategy has to do with the processing of ores into the products we consume. Many of the toxic chemicals that are released into the environment today result from the transformation of the minerals we mine and their manufacturing into finished goods. A range of compounds—*intermediate chemicals*—are produced for those purposes. Sometimes, toxic metals are used in the same way and are part or by-products of the fabrication process. Once used, they often end up in lagoons, waterways, and the atmosphere.

For example, mercury was once utilized for the extraction of gold and has been used in the fabrication of paper and plastics. It is also contained in the ores of other minerals, the smelting of which releases the volatile metal into the atmosphere. The burning of coal for electricity or other purposes also results in the emission of mercury into the air. That is, in addition to all the other pollutants its combustion produces.

By virtue of the fact that nothing is created nor destroyed, intermediate chemicals also end up in the environment, in their original forms or as byproducts.

The second principle of a comprehensive environmental strategy is that cutting down on the extraction of minerals would not only preserve resources and reduce pollution directly but also achieve a third goal: decrease our use of the invisible battery of toxic compounds produced for the processing of ores and manufacturing of goods.

## The Third Principle

There is an additional principle in a comprehensive environmental strategy: the greater the world population, the more resources are consumed at any given point in time. For example, if the total number of people on the planet were reduced by 30% and everything else remained the same, we would consume 30% fewer goods and use up 30% fewer resources. We would need to extract 30% less mineral, burn 30% less fossil fuels. We would create 30% less waste, pollution, and environmental degradation.

The real long-term foundation of a sustainable economy, of a green society, is resource conservation. Today's efforts towards the environment only skim the surface and have no hope of ever catching up to the current rate of environmental destruction. There is just too much being dumped into the environment, from the billions of tons of minerals extracted every year, to the millions of gallons of intermediate chemicals, to the multitudes of soaps, detergents, solvents, and cosmetics flushed down rivers on a daily basis.

The third principle of a comprehensive environmental strategy is that we have to reduce world population. Even the greenest measures and processes will not be able to compensate for the devastation that billions of people can wreak upon the earth.

Efforts to develop cleaner and greener processes and products need to continue, but they are not enough. We need a much more profound and powerful mechanism to turn things around for the environment. We need to better understand economics and use markets to our advantage in order to drive the political agenda. We need to create a *green economic environment*, a marketplace in which environmental practices are profitable and their opposites are not.

Next, comes the piece de resistance of *The 21ˢᵗ Century Environmental Revolution*. We will look into more details at the sources of funding that could support a large-scale environmental strategy and the engine of change that could create a green economic environment.

## Funding for a Comprehensive Environmental Strategy

We are not doing enough for the environment. What is needed is a totally new approach capable of handling 80% to 90% of the problems. What is needed is massive change. That means, moving to larger scale strategies and more fundamental approaches.

Really turning things around for the environment could cost trillions of dollars for the US and Canada alone. Moreover, that kind of money would be needed on an annual basis. Let's look at the options.

### *Taxation*

Funding has always been the stumbling block of the environmental movement. Fixing problems costs money. This is why our efforts are still only skimming the surface.

People often complain of giving too much of their hard-earned cash to governments. Doubling taxes to raise the trillions needed for the environment is not going to happen. The countries that do tax for the environment have only been able to raise minimal amounts for that purpose. There is no particular support, either on the part of politicians or voters, for any large amount of additional taxation. The most charismatic environmentalists already have enormous difficulties raising money for the little we currently do.

As it stands, additional taxation can only have a minimal impact for the environment because of its hopelessly insufficient fundraising capability.

### *Subsidies and Tax Breaks*

When governments give grants or tax breaks to corporations, they have to make up for the shortfall in revenue through additional income tax, cuts in services, or other similar actions. Subsidies and tax breaks are not as free as many would think; taxpayers ultimately foot the bill. As such, these approaches to resolving environmental problems are as limited as additional taxation is.

## Regulations

Regulations have scored a number of victories for the environment, for example, outlawing PCBs and chlorofluorocarbons (the compounds responsible for ozone depletion). However, they are also ultimately limited by the availability of funds. PCBs are still in use today in closed systems. CFCs will still be legal for a number of years in developing countries. Why is that?

Regulations do cost money either in the form of higher prices for alternative products or because of the costs of conversion. These are usually passed on to consumers. Once again we pay the bill, this time in the form of more expensive retail products. This is why regulations are often much weaker than needed, assuming they exist and are enforced.

Short of people dying, there is a limited amount of social commitment to seriously address environmental problems and even less with respect to long-term preemptive and proactive planning. Regulations will remain an important tool, but alone they have no hope of ever solving our problems. We may be willing to pay a little more for greener alternatives but not what would be required for an environmental revolution.

## Policies for the Future

Regulations have decreased the number and amount of contaminants being dispersed into the environment. However, because of their inherent limitations, they are only touching the surface.

Resource conservation has generally failed to make it to the social and political agendas. Most often, the only resources targeted with our efforts towards sustainability are forestry and fisheries. This is no great achievement as both are actually renewable. Despite that fact, we still managed to fail. The collapse of the cod fishery on the Grand Banks on the eastern coast of Canada and the US is but a striking example of our impotence and lack of resolve.

We have not even touched the issue of non-renewable resources yet, which is by far much more critical. The world is again in a fossil energy crisis, hoping that OPEC increase, not decrease, production.

What does that do to our Kyoto targets? Nothing is done to conserve metals. On the contrary, calls are being made to increase production in order to create jobs.

Sustainable development runs over and over again into the same brick wall: *counter economics*. The very incentive structure of the system in which we live directly pushes us to deplete non-renewable resources to keep prices down or create jobs. Similarly, it is cheaper to pollute, use toxic compounds in the manufacturing of goods, and flush hazardous waste into waterways than opt for cleaner alternatives.

Today's reality is that the more we pollute and decimate resources and the less we leave our children, the greater our economic growth. Two things are needed to address environmental problems. Firstly, the solutions have to be on the scale of the problems. Secondly, the current economic structure has to change. No matter how hard we fight for the environment, if money runs counter to our efforts, little will be achieved. How do we do that? Let us first look at what has been proposed before.

## Hitting The Jack Pot

The idea of environmental taxes is not new. In 1989, David Pearce produced for the Department of the Environment in Britain a document which recommended a comprehensive green taxation program in the UK (Dryzek, 2005, p. 130-132). Canada, the US, and many European countries have already implemented some form of environmental levies, all be it very minimal. The global warming debate has led to talks of carbon taxes.

Could trillions of dollars in additional taxation for the environment solve our problems? Probably. The scale would be appropriate, but the idea would not be very politically viable. Could we not find trillions' worth of spare change within the existing taxation system? The answer is also very likely *no*. But...

Many countries have deterrence or punitive taxes such as levies on cigarettes and alcohol. These are meant to do two things: deter usage and generate income for the government. They are *dual-purpose*

levies.   When governments tax income, they bring in revenue but
nothing is specifically being deterred.   Because this is a *single-purpose* type of levy, society derives from it only one benefit when it
could have had two.

Shifting taxation from income to the environment would double
the social dividends we get from our tax dollars.  This would create
synergies that could achieve a lot while not costing us a cent.  There
might not be trillions of dollars of spare change available within the
current taxation system, but there definitely is that amount of wasted
incentive or social benefit.

Considering that Canadians and Americans alone are paying about
U.S. $2 trillion annually in income tax, the benefits of such a strategy
could be massive and spark an environmental revolution within half a
decade. Let's take a closer look.

## A New Budget for the Environment

There is no doubt that levies have to be collected to fill governments'
coffers and fund social services, healthcare, education, roads, the
police, the army, etc.  But, who is to say that income has to be taxed
in order to raise the money for the above.  Canadian taxation represents over 30% of national income.  US taxation is somewhat lower
but remains a massive component of the country's revenue.  Those
represent staggering amounts.

Let me illustrate with a concrete example what a shift in taxation
could do.   Income tax is a single-purpose levy designed to raise
revenue for governments.  $100 of it produces $100 worth of services.

With deterrence taxes, governments collect revenues to provide
services as well as deter the use of products which are generally
considered unhealthy.  $100 of those provides not only $100 worth of
services but also $100 of deterrence.  In the case of cigarettes, the
latter could translate, for example, into lower healthcare costs from
decreased cancer rates.  In total society would derive about $200
worth of benefits.  Taxing income in comparison is very wasteful,
providing only half as much.

In North America, when you tax income—as opposed to goods
like cigarettes—you get U.S. $2 trillion into the government coffers,
but you lose another U.S. $2 trillion worth of deterrence effect.  That
alone could rapidly bring about an environmental revolution without

costing the taxpayer a cent, i.e. without additional taxation!

Most resource conservation and environmental issues are ones of deterrence. We want to deter the use of pollutants and non-renewable resources. Shifting the single-purpose income tax burden to an environmental deterrence tax system would double the social benefits we get from every dollar of tax paid without increasing taxation. The U.S. $2 trillion of deterrence that this would generate for North America alone would be a massive engine of change for the environment.

Of course, our lifestyles would change, but that is unavoidable. We cannot achieve an environmental society, a greener world, without anything changing. But our standard of living would not go down. Our disposable income would essentially remain the same, the total amount of taxes paid generally being equal before and after the shift from single to dual taxation.

We already know that there is no social commitment to any significant increase in taxation and that the current amount dedicated to the environment is far from adequate. The shift to deterrence taxation would allow us to get a second bang for free from an existing tax. If today you are paying $10,000 in income tax, tomorrow you would still be paying about $10,000 in taxes.

However, the latter would not come from income but from deterrence levies, for example on contaminants and non-renewable resources. $10,000 worth of revenue would still be generated for the government, but the punitive effect would reduce pollution and foster resource conservation without your paying any more taxes than before.

In 2000, Canada collected on incomes about Can. $127 billion (Revenue Canada, 2002). Americans paid U.S. $2,098 billion in taxes in total during the same year. Assuming that the currencies are at par (Can. $1.00 = U.S. $1.00), this would amount to about U.S. $2,225 billion (or $2.225 trillion) for North America.

In comparison, the US gross national product (GNP)—a measure of the total value of goods and services produced by a country—for 2000 was about U.S. $10 trillion. The actual annual environmental deterrence budget—the *green firepower*—for the US alone would be about 20% of its entire national production.

Internationally, the story would be similar. The rest of the developed world would have a comparable share of its total production available for environmental deterrence. Developing countries, which have been able to afford environmental standards only with difficulty up until now, would have less but still enormous amounts of deterrence at their disposals. By any standards, we are talking about massive and unprecedented firepower for the environment in *both* rich and poor countries.

The GNP is very effective in giving us a good idea of the actual size of the new potential environmental deterrence budget. But, a much more interesting comparison is that with the American military budget. It stood at U.S. $379 billion for 2003 (Council for a Livable World), in the midst of the Afghan and Iraq Wars. At $2,225 billion, the new budget for the environment would not only beat but actually dwarf the war chest of the most powerful military on the planet.

With this kind of budget, an environmental revolution early in the 21st century becomes more than a mere possibility. It is right in our hands, just waiting to happen.

# 7. Premises of Large-Scale Environmental Change

Large-scale change is by no means something easy to pull off. Most deem it impossible. However, if certain principles are respected, it can happen. There is a science to massive change.

## The First Premise: Massive Scale

The first premise of a large-scale environmental strategy is that needs are massive. Therefore, the solution we should be looking for should also be massive. There is an urgency to deal with the problems of pollution and environmental degradation. There is also a dire need to reverse the current trend of rapid resource depletion so that future generations can have enough for themselves. The sad reality is that the earth does not come with a warranty stating that because its life expectancy is five billion years, its resources should last that long. The mechanism we are looking for needs to address both issues.

The small step by small step incremental solutions proposed so far will not do the job.

## The Second Premise: Strong Political Support

The second premise is that large-scale environmental change requires massive voter support. For any solution to be implemented, it must first make it to the political agenda and receive broad support. To that purpose, the solutions explored in this book are as voter friendly and as politically viable as possible given the ambitious task at hand. It is my hope that the case I make is also economically solid enough to get experts on side and quickly bring the strategy proposed to the political arena.

Because society's commitment to the environment is ultimately limited, large-scale change would have to be achieved with virtually no additional amount of it. This brings us to the next premise.

## The Third Premise: Minimal Social Commitment

The third premise is that to be voter friendly and get massive support, strategies for large-scale environmental change have to be able to be implemented essentially without additional social commitment.

Current solutions always run into the same obstacle: voters do not want any additional taxes, at least not on the scale required. Designing strategies based on full social commitment is unrealistic. We need solutions that are able to address 80% to 100% of environmental problems with roughly the same amount of social commitment as we have today.

## The Fourth Premise:  Implementability

The fourth premise is that large-scale environmental change calls for practical and implementable solutions. Remember Keynes' famous quote, "Foul is useful and fair is not"? Rather than an attempt at cynicism, it was an acknowledgment of the basic reality of economics. Theories and principles are not very useful if they are not applicable. If he had spoken about the environment, he might have said, "What does not make it to the political agenda and sway the voter has very little use."

Much of the money spent on research every year does not benefit society for a number of reasons. Studies often do not result in practical applications, they are never brought to the general public's attention, or the solutions proposed are just too unfriendly to voters to be politically viable. Of course, not all research is meant to produce solutions. Some types are supportive in nature.

*The 21ˢᵗ Century Environmental Revolution* was written strictly from an implementability point of view. What is not implementable and politically viable has very little use. The mechanism of change described in the next chapters is as practical and voter friendly as it can be, considering the fact that we are contemplating massive change.

## The Fifth Premise: Market Efficient Mechanisms

The fifth premise is that, to maximize implementability and political support—especially given the magnitude of change we are considering—the environmental solutions proposed should make use of the most efficient mechanisms available. Let's review the options at our disposition.

Current research divides approaches into two main categories, regulations (*command-and-control*) and taxation. Regulations can be very effective in specific cases and have their place on the environmental agenda. However, they generally lack efficiency on several counts.

Ian Bailey (Senior Lecturer in the School of Geography department, University of Plymouth, UK) reports that they are viewed as economically inefficient because they are imposed uniformly on polluters who have different abilities to adapt to better practices (Bailey, 2002, p. 235). For example, big corporations can more easily invest in cleaner technologies than smaller businesses.

Annegrete Bruvoll (1998) further argues:

> The required control costs under regulations can be considerable and the administrative costs are usually higher than under taxes. Also, price incentives offer free cost information, while mandated recycling is often accompanied by expensive education and motivation programs. (p. 19)

In other words, regulations often have a number of costs associated with them—administrative, implementational, educational, etc. The same author also reports that studies had consistently shown regulatory programs to be poorly designed and have costs exceeding benefits.

Another problem is that, unlike taxation—which is more efficient and promotes continued improvement—regulations are not *dynamically* efficient. That is, they offer little incentive for polluters to exceed minimum requirements and do better year after year. In addition, without international agreements to back them up, regulations are further limited as they tend to impede the competitiveness of host countries.

For these reasons, the focus of policy has recently been shifting to market-based solutions. These new approaches are increasingly viewed as the best ways to address environmental issues. Bailey (2002) states that "both theoretical and empirical studies suggest that market-based mechanisms remain the most efficient method for pursuing many environmental goals, so that countries pursuing such strategies generally perform better economically than those that rely on legislative standards" (p. 237).

### Market-Based Options

Bailey (2002) describes three main market-based options, *incentive taxes*, *cost-covering charges*, and *revenue-raising taxes*. Incentive taxes are designed to increase the costs of polluting in order to bring about positive environmental behaviors. For example, toxic chemical compounds or fossil fuels can be taxed to reduce their use.

Cost-covering charges are usually levied from users of a particular resource and used to manage it or mitigate its degradation. For example, entrance fees to national parks may be used for their maintenance. Pollutants might be taxed to raise funds to undo the damage they cause.

There are two types of revenue-raising taxes. The first, *additional taxation*, involves for example environmental charges on plastic soft-drink bottles or residential tipping fees on garbage. The revenue generated through these go directly to general government tax coffers rather than being targeted for specific purposes. They are additional charges to consumers and businesses (which pass those on to us) just like the first two types of market-based approaches discussed above.

As such, all three options above are limited to the current minimal level of social commitment and, therefore, to skimming the surface of environmental problems. This brings us to the last option.

The second type of revenue-raising options is revenue-neutral taxation. Environmental levies are collected and offset by lowering other taxes, for example, on income or retail sales. They do not increase the general level of taxation for people in a country. They are only a shift from one form of levies to another.

For example, if a government were to raise $1 billion in new environmental taxes, it would decrease income or retail taxes for taxpayers by the exact same amount. A government would take with one hand and give the same amount back with the other. Because environmental taxation is offset by a drop in other taxes, it would be revenue neutral. The levies would result in a deterrence of unenvironmental behaviors but come at no cost to the taxpayer.

The key phrase here is *at no cost to the taxpayer*. This essentially gives us non only an efficient mechanism but also a politically-viable one. It is the massive engine of change that could bring about an environmental revolution.

## The Sixth Premise: Focus on Resource Conservation

The sixth premise is that the primary focus of environmental policy should be resource conservation. Pollution and contamination are in most cases easier to prevent than clean up. To what extent can gases from industrial processes or the incineration of garbage be cleaned up? Bruvoll (1998) writes, "Most [waste] materials are dispersed and hard to collect after use at the emission stage, and designing and operating a tax system that tracks all environmental effects is out of reach" (p. 16).

Resource conservation is therefore a better primary focus for overall environmental policy as it attacks the problem at the source through prevention, not only preserving resources for future generations but also reducing pollution, intermediate chemical use, and environmental degradation from waste disposal. In addition, making it the primary focus of policy forces us to address non-renewable resource depletion together with environmental degradation. Current policies lead us to give little or no attention to preserving resources for future generations.

Of course, the above does not mean that other issues will take a back-seat to conservation, on the contrary.

## The Seventh Premise: Landfill Sites Are Limited

The seventh premise is that landfill sites are also a limited resource. There is not an inexhaustible supply of suitable sites to dispose of our garbage. Because of the toxicity of some materials and concerns about seepage of contaminants into water tables, not all locations are geologically suitable for waste disposal.

For economic efficiency, landfill sites are often located in close proximity to towns, threatening the contamination of drinking water sources. Current recycling efforts only partly reduce the continued demand for landfill space. They do not eliminate it by a long shot.

# 8. *The Massive Engine of Change*

## The Current Incentive Structure

Under normal circumstances, you would expect governments, and certainly more so an economic system, to reward good behaviors. Not doing so could lead to a very dysfunctional society. Rewarding or failing to discourage undesirable behaviors can create very disruptive and destructive results. Profits and taxation are perhaps the two most important incentive components of the economic system.

The environmental mess we are currently in is the direct result of an incentive structure that turns a blind eye to the excessive use of pollutants, production of greenhouse gases, consumption of non-renewable resources, etc. The way the current economic system is set up often makes it more profitable to use cheaper and more toxic compounds. The decimation of non-renewable resources means job creation and great profits. In other words, current economics actually rewards the wanton destruction of the environment and resources. Why is it so surprising that the planet is in a total mess?

The market system on which the entire world economy depends is configured into a destructive mode.

## The Environmental Taxation System

The only approach capable of bringing change on the scale that is needed to turn things around for the environment would have to involve markets. One way to do this is to change the incentive structure that drives them. The easiest and most obvious means to achieve that is to modify taxation, which is what is being proposed here. The *Environmental Taxation System* (ETS) is a strategy that would reduce income and retail taxes in favor of environmental levies on non-renewable resources, pollutants, and fossil fuels. The changes would be revenue neutral, i.e. overall taxation would not increase.

## A Brief Overview of the ETS

Our current taxation system can easily be characterized as a politically-driven shotgun approach to filling the tax coffers. It is dysfunctional economically and disastrous for the environment. At this point, not much thought is being given to the fact that taxation is the very incentive structure that shapes our lives and the economy of our countries.

The lack of emphasis on the environmental component of taxation leads to behaviors that make us deplete and waste non-renewable resources and pollute the planet as if there were no tomorrow. Current environmental problems are in large part caused by the improper structuring of the economic system in which we live.

That can be changed by altering the tax incentive structure to reward conservationist and environmental behaviors and punish their opposites. Under such a system, government revenues would come primarily from taxes on non-renewable resources, contaminants, and fossil fuels.

As the ETS would be revenue neutral, a government would collect the same total amount of taxes after its implementation, but it would come from different sources. For example, if 60% of taxes came from income, 38% from retail sales, and 2% from the environment, the shift could result in a new distribution looking like this: 40% coming from income, 10% from sales, and 50% from the environment.

The new incentive structure would create a green economic environment and new consumer landscape in which the price of non-renewable resources, pollutants, and all the products made with them would increase. The goods made with renewable and non-toxic materials—such as wood and reused or recycled materials—would become cheaper or gain a competitive advantage.

As a result, consumption would naturally and effortlessly shift to greener products, behaviors, and sectors of the economy. Good citizenship would be rewarded rather than punished, and the destruction of the planet would be deterred rather than encouraged as is currently the case. The ETS would create a *greening effect* that would make the world more environment-friendly year after year.

As a general rule, the ETS would tax non-renewable resources—such as metals—at the source to promote conservation and create natural markets for recyclables. These would replace the current municipal recycling programs funded by taxpayers. Simple higher input costs would also be more easily understood and managed by businesses compared to a haphazard collection of regulations.

Substances that are damaging to the environment (toxic compounds, carcinogens, etc.) would be taxed, whether they are inputs in manufacturing processes or consumer goods.

Oil and other fossil fuels would be taxed not only to decrease usage and greenhouse gases but also to create a stable and profitable renewable energy sector. This would replace the existing grant and tax-break incentive system, which is inefficient, costly, and creates distortions in the economy.

Regulations would be used in the packaging industry. In conjunction with ETS taxes, they would create natural markets for container reuse. This would serve to replace most of the existing refundable-deposit system and result in efficiencies and a drop in costs for the consumer.

The shift to environmental taxation would be fully scalable. That is, we could implement the ETS as fast or as slowly as we want. This would enable us to make the transition to the ETS as quickly as possible without creating economic havoc or unnecessary human suffering. Scalability would allow us to start with lower taxation levels—giving us and the industry time to adjust—and progressively increase them after. It would also make it possible for countries to give different weights to each of the components of the ETS according to their own goals and political imperatives.

For companies, the ETS environment would make it more profitable to preserve rather than waste resources. The opposite would be true for the use of toxic materials and contaminants in the production process. The green economic environment would deter fossil fuel use and promote alternative energies within the existing economic system.

## High-Efficiency Systems

Many environmental policies are inefficient and expensive to administer. They are cumbersome to the business sector as they require that staff keep up to date on regulations. It is usually much easier for companies to make decisions based on the price of the resources needed to manufacture a product than have to deal with a maze of government rules. Input cost is a language more easily understood and simpler to use in planning and decision-making. It is already part of normal business practices and expertise. This is the language that the ETS speaks.

Environmental taxation would raise the cost of an input in a predictable way. Businesses would make decisions on that simple parameter. Current environmental policy is cumbersome to governments because regulations need to be monitored and enforced, often necessitating additional bureaucracy.

A second aspect of efficiency has to do with the total number of tax collection points. That is, the total number of individuals and businesses that governments collect from. Removing part of the taxes people pay on income would eliminate the need for many individuals to submit tax returns or, at least, would make those much simpler. That could mean thousands fewer or easier filings to be processed.

However, much more interesting is the effect on retail levies. Many governments currently collect retail taxes from just about every single business selling goods or providing services to consumers. This represents thousands of corporate taxpayers and means a lot of unnecessary bureaucracy.

As you go up the production chain—from resource extractors, to input producers, to manufacturers, to retailers—the number of individuals from whom governments have to collect taxes grows. For example, a copper mining company might sell its products to 10 part makers. Each of these in turn might sell parts to 10 manufacturers. At this level, 100 businesses would have to be collected from. Each of them might sell to 10 retailers. Collecting taxes at that level would then involve 1000 companies. In other words, the higher up you collect, the more expensive it is not only for governments but also for businesses and the economy as a whole.

The ETS would collect taxes mostly at primary levels, therefore minimizing the number of players involved. Furthermore, it would do so only from target items: non-renewable resources, contaminants, fossil fuels, etc. Green sectors would generally see taxes disappear from their products.

### *Dual-Level Planning*

There are two main levels of planning in the environmental arena, the general (or macro) and intricate (or micro) levels. Governments are responsible for planning at the general level. This is where we state what we want and establish plans of action to achieve our goals. The difficulty with environmental issues is that as a situation increases in complexity—for example, in day-to-day business operations and the multiplicity of situations encountered in reality—it becomes more and more difficult to manage. Some problems are just too complex and intricate even for the best specialists in governments.

It is not really a matter of failure. For example, even with the best technology available to date, we cannot predict the weather with any degree of accuracy for more than a few days ahead of time. It is one of proper management. Currently, government regulations and programs are the main planning mechanisms used to address environmental issues. They fail to do the job. Furthermore, when they do not work, the taxpayer ends up footing the bill. We have to look for a mechanism that will have the appropriate expertise at the micro level while working within general government guidelines.

The market has a multitude of agents—business people—each an expert in the field. Nobody else can better estimate the effect of higher input costs on the final price of a product or the bottom line. All planning on their part is based on familiar market decisions and comes free of charge to the taxpayer. As well, any error results in losses to them, not the taxpayer. The private sector represents a massive wealth of expertise at the micro level. If unqualified for planning at the general level, the business community possesses the detailed knowledge necessary to make the best economic decisions in regard to their own markets. Governments cannot hope to approach that kind of expertise.

The ETS would rely on governments for general directions and on the market for complex decisions. This would provide for both types of planning. Setting tax levels would enable governments to guide the economy but allow the market to make decisions in the best possible and most efficient ways. The strategy would lead to a much more effective and appropriate combination of planning and expertise, and maximize change and benefits to all involved.

### Business-Friendly Green Policy

The ETS would reduce the regulation and tax collection burden on industry. It would also provide for much simpler decision-making for businesses. The fact that the system would be highly flexible, fully scalable, and comprise several components that could be implemented independently would allow for a smoother implementation for everybody.

### New Approaches to World Development

A worldwide implementation of the ETS would give rise to several new mechanisms for world development. Since the 1970s, the developing world has been hit by one oil crisis after another, preventing many countries from gaining significant ground on poverty. As renewable energies are diversified and can be produced nationally, developing countries would benefit from the ETS in terms of balance sheet and job creation.

Global resource conservation would increase the international price of most metals as well as of the goods made from them. Developing countries would get better prices for their commodities but could still choose lower ETS tax levels to increase their competitiveness and gain market shares, boosting their own economy and promoting their industrial development. Taxes could potentially be waived for resource-poor developing countries.

Environmental regulations have always been an issue of contention in international markets. The developing world has historically been perceived as having an unfair competitive advantage over developed countries because of lower environmental standards. We do not need to be caught in a race to the bottom.

Comprehensive ETS-based international agreements could raise standards across the board to benefit the entire world community while not affecting competitiveness. Conversely, a unilateral increase of environmental standards in the developed world could be used as a means to increase the competitive advantage of developing countries. The former would benefit from a cleaner environment while the latter would get an economic boost from a greater share of markets.

### *Tax Fraud*

Most social and economic systems are prone to fraud to various extents. The fact that the ETS would be levied at the manufacturer level would significantly decrease the number of collection points. As such, the system would be not only more efficient but also less prone to fraud than current widespread retail taxes.

Black market or illegal work is a problem around the world. The bill for the *underground economy* can add up to hundreds of millions of dollars every year. In Canada, people do not pay income tax on their first $10,000 worth of revenue (the basic deduction). As such, black-market fraud up to that amount is mostly meaningless; no one pays income tax on the revenue anyway. The ETS would shift the first tier of income tax (for example, $30,000 in Canada) over to the environment. This would mean that the first $40,000 dollars of a person's income would no longer be taxable and subject to fraud. This could have a significant impact on government revenue and result in lower overall taxes for the rest of us.

## A New Future

The Environmental Taxation System is the massive engine that could bring about the large-scale change that is needed for the environment. Switching to this new incentive structure would create a green economic environment that would turn the current market system from a destructive power into a powerful force for positive change.

The switch to environmental taxation would take advantage of the dual effect of levies—one that is wasted by income and indiscriminate retail sales taxes. At no overall additional cost to taxpayers, it would gently but radically transform the economic system in which we live into a mean lean green machine. It would make environmental change possible to occur at a speed unanticipated and not even dreamed of.

The implementation of the ETS would change the incentive structure of capitalism from a system that essentially punishes people for working (income tax) and consuming (sales taxes) to one that promotes resource conservation and environmental responsibility. It would transform a system that wantonly destroys resources and the environment into one in which businesses would have a vested interest in preserving them both, not the opposite as is currently the case. This new overall incentive structure would foster and reward conservationist and environmental habits in both the industry and consumer markets.

The ETS would not only be massive as an engine of change but also highly efficient. Its administration would be much less costly than our current patchwork of regulations as it would make use of an already well-understood and efficient mechanism: taxation.

Furthermore, the ETS would speak a universal language, the language understood by workers, business people, consumers, and voters: money. Once the new rules are set, societies would progressively refashion themselves around green economics and environmental lifestyles.

The scalability of the ETS would allow for choice in speed of implementation and help soften its impact on particular economic sectors. Its three components—non-renewable resources, contaminants, and fossil fuels—could be implemented together or independently as desired.

Unlike many other alternatives, this new approach to environmental issues would not be anti-business or anti-economic. It would work in concert with the main components of the current market system. Overall, businesses would not be stuck with the costs of this green transformation anymore than taxpayers. On the contrary, the ETS would make it profitable for companies to invest in green practices and environmental research and development (ER&D).

Many business opportunities would arise following the implementation of this new tax structure. The current catering to niche green markets would become a stampede for opportunities in an exploding new economic frontier. Countries and corporations at the forefront of ER&D would find themselves in a race to capture the new 21$^{st}$ century world markets. Wasters, polluters, and those who do not care for the environment or are unable to adapt would only fall behind and see the fate of the dinosaurs.

A comprehensive environmental tax strategy would give rise to a new overall mechanism for world development. It could allow for a partial shift of the world industrial base to developing countries.

## General Issues Relating to the ETS

Governments collect taxes from a number of sources, for example, income, retail sales, the environment, and profits. In addition, specific products such as cigarettes and alcohol are often taxed for deterrence purposes.

For the sake of simplicity, the following discussion will concern mainly the first three types. Of those, income is probably the main source of revenue for most governments, with sales coming in second and environmental taxation being minimal. The ETS would progressively shift the tax burden from income and sales to the environment.

### *Keeping Taxation Progressive*

Most Western countries have a progressive taxation system. That is, people contribute according to their ability to pay. For example, in Canada the tax rate on the first $30,000 of net income is about 30%. It is the first bracket. If that were the only one, we would have a flat tax system. That is, everybody would pay the same rate, no matter the income level. This first bracket is actually a flat component in a progressive tax system.

As one's net income rises above that, a higher rate is paid. For example, Canadians are taxed at 45% in the second $30,000 bracket. For the next two tiers above that, we pay respectively 55% and 60%. In order to keep taxation progressive under an ETS system, only the flat portion of income tax would be suitable for a shift to the environment. In Canada, that would be the first bracket, or the tax on the first $30,000 of net income after the basic deduction.

In addition, we also have to be careful with respect to how we go about eliminating that bracket. Simply increasing the basic deduction (about \$10,000 in Canada at the moment) to reduce the total taxable income would be regressive because it would lower taxable incomes at the top where people pay the highest tax rate. That would be regressively cutting back taxes.

If a government chose to increase the basic personal deduction as a means of implementing an ETS system, the tax bracket thresholds would need to be adjusted lower by the same amount to avoid worsening the redistribution of income. Alternatively, a government may simply set the tax rate in the first bracket to 0%.

The ETS would imply a large reduction in income tax for everybody. We have to make sure that it is done in a way that maintains the status quo with respect to progressiveness.

Most countries have retail sales taxes. This is what you pay on top of the purchase price of items that you buy in stores. In Canada, that would be the Goods and Services Tax (GST). This type of sales levy is not progressive as the same rate is charged equally to everybody. As it is a flat tax, it would fully qualify for a shift to environmental taxation. Any state or provincial sales taxes of the kind would be equally suitable for the shift.

### Compensation for Lower Income Earners

The increase in environmental levies would have to be compensated for in social support programs as their recipients generally do not pay tax on the financial assistance they receive and would see their real taxation increase under an ETS shift. Lower income earners may face a similar problem because their basic deductions are large compared to their taxable income. Governments may have to look into making appropriate adjustments to the system so that their total taxation remains the same and the shift to the ETS is fair to all.

In the late 1980s, Canada went through a tax shift. The manufacturers' sales tax—a hidden levy collected at the manufacturing level— was replaced by the GST, a retail tax. Upon the change, the federal government opted to compensate lower income Canadians by issuing checks on a quarterly basis to millions of people in the country. It is still writing these to this day, some 15 years later. This is not exactly the most efficient way to do things. Adjustments need to be made directly to social support programs and the taxation system so that the actual benefits received or take-home pay compensates directly for the ETS.

### Revenue Neutrality and Transparency

The transition to a green economic environment needs to be as voter friendly as possible. To that purpose, the cornerstone of the ETS is that it is revenue neutral. For every green dollar collected, taxpayers would see their income or sales taxes reduced by one dollar. On average, the same total amount of taxes (income, sales, and environmental) would be paid by taxpayers before and after the implementation of an ETS system.

Revenue neutrality is a fundamental aspect of the ETS and is key to its political viability and implementability. Although voters may be willing to pay taxes differently, they would not be so easily convinced to give more. The ETS absolutely needs to not increase total taxation. To ensure proper and continued political support, governments would have to be highly transparent with respect to revenue neutrality and report on it regularly.

# 9. Implementation Issues and Scenarios

In the next few chapters, we will be moving from general principles to the specifics of implementation. Concrete examples will be looked at in order to demonstrate the feasibility of the ETS and dispel misconceptions. For many socio-economic issues, much of the resistance to change comes from unfamiliarity. Seeing environmental taxation from various angles and in real situations should get us used to the idea that the transition to the ETS would be much easier than it appears to.

## The Silent Scenario

This scenario will serve as a benchmark to assess to what extent the ETS would help achieve our goals for the environment. First, let us remove the financial aspect from the debate and see what we need to achieve for the environment. Let us silence the money concerns and assume that we are in a perfectly planned world. You are the head of state of your country and have to draw a comprehensive plan to address environmental issues.

To conserve non-renewable resources for future generations, you would need to reduce their extraction. This would mean that the mining industry would decrease in size. Unfortunately, this is a necessary evil; you cannot conserve resources and keep digging them out of the ground as fast as you did before.

It would also mean that we would need to consume less of them. Aluminum soft-drink containers would have to become reusable or be replaced by glass. In many cases, tin cans for food storage would disappear from store shelves and be replaced by environmentally-friendly alternatives. Metal components in various products would have to be more strategically used and substituted for whenever possible.

Metal-heavy sectors would be hardest hit under the ETS and see the most transformations. The automobile industry would have to build vehicles that are conservational, i.e. that use non-renewable resources sparingly. That could be achieved by replacing metals with substitutes wherever possible and decreasing car sizes.

Automobiles would have to be kept many years longer and repaired more extensively. The used car industry would thrive as people would buy fewer new automobiles and get them fixed more extensively as a means to conserve metals. The production of new vehicles would decrease slowly in the short term as new generations of conservational, fuel-efficient, and alternative-energy cars and trucks would be needed. In the medium and long term, the industry would progressively shrink to a smaller size.

Automobiles and other vehicles would need to become more fuel efficient as well as switch to alternative energies. Either way, we would consume less gasoline. We would pump less oil from wells and have to start producing renewable fuels.

New jobs would be created in alternative sectors in general as the demand for their products rises. The wood and renewable resources sectors would thrive and need to be properly managed to prevent their collapse as has already happened in some cases.

### Feasibility

One of the main concerns with respect to sweeping economic changes is the uncertainty they can create. Economists are still looking for the perfect theoretical solution to environmental problems, but there can be a huge gap between theory and practice, especially if the changes involved are of significant magnitude.

Brand new, maverick solutions are likely to face a lot of resistance especially if they have never been tried in the real world before. They may never be implemented simply because of the large amount of uncertainty they involve. This is a significant concern when searching for answers to environmental problems: what is not implementable because of lack of political viability is not useful.

Although the ETS is new and sweeping in magnitude, it relies on an old practice: taxation. The new system would only be a shift from one existing component of the system to another. Environmental and deterrence taxation are already in place in many countries. Along with income and retail sales taxes, they are all components of total taxation.

As expressed earlier, in the late 1980s Canada had a major shift in taxation. It eliminated the manufacturers' sales tax and replaced it by a general goods and services tax collected from retailers. The new levy was about 7% and applied to most goods and services sold in the country. For the first time, services were taxed in Canada. The transition went relatively smoothly despite being met by initial distrust and opposition. As such, we know that a taxation shift can be done.

An ETS shift would be revenue neutral and should not change the overall amount of taxation collected in a country. For this reason, the new strategy would not put recessionary pressures on an economy. Consumer spending on goods and services would remain essentially the same.

For example, with the ETS people in a typical country might pay $1 billion more in environmental taxes, but their income and retail taxes would be reduced by the same amount. As a result, their purchasing power would remain the same. In other words, many things would be more expensive, but consumers would have more money to spend and come out even in the end. As long as the consumption level remains the same—which would be the case with the ETS—there should not be a reduction in economic growth or significant net job losses.

Many environmentalists see consumption itself as *the* problem. They are correct to some extent, but there is more than one way to reduce our environmental impact: we can decrease consumption, but we can also make it greener and achieve the same result. Opting for the former would mean a slowdown in economic growth, and few are ready to support that at the moment. The option would simply not be politically viable.

The ETS would progressively make existing levels of consumption greener and greener without threatening to send economies into recession. For that reason, it is currently the best viable large-scale option we have. Of course, it is a compromise but a very good one. Under its spell, economic growth would come mainly from green sectors. Unenvironmental industries and the consumption of their goods would shrink and certainly not grow.

The green sector is an emerging market in which there is not a huge amount of competition at the moment. Implementing the ETS in a country would give it a significant competitive advantage in what will be a strong growth sector of the future. Export markets are likely to grow and result in a lot of job creation.

The ETS is essentially based on established and well understood economic practices. It is readily implementable and has a relatively low level of uncertainty, given the magnitude of the changes that would result from its implementation.

## Geopolitical Concerns

The ups and downs in the price of oil over the last few decades have given us some insights on the geopolitical implications that the depletion of resources can have for modern economies. Decreasing reserves can be very disruptive. The conditions created by shrinking supplies increase the potential for deadly conflict. Scarcity is a very bad thing.

A global move to resource conservation could significantly extend reserves and slow down the depletion process, easing up the pressure for potentially chaotic events in the future. Mineral resources other than oil have been less vital to most economies because their supply has remained largely uncartelized and unpoliticized. They are also less massively used compared to energy.

However, metals are much less replaceable than oil. Alternatives to petroleum do exist and are actually plentiful. They are just more expensive. That is not the case for other mineral resources. What will happen when their supplies begin shrinking like petroleum? They are a major component of the infrastructure of most countries, and modern lifestyles depend on them.

Without conservation and at the current rate of depletion, shortages will eventually begin to occur, and the economics of metals could resemble those of oil. Power could shift to mineral-rich countries just as it shifted to the Middle East with energy. More likely, as metals are more widely distributed than oil, countries which import much of their mineral resources may see power shift away from them, their wealth following suit. That could be the case for the US. The question is, do we want to accelerate this process with high rates of depletion and precipitate more global crises, or do we want to conserve resources worldwide and decrease the potential for problems developing as a result of scarcity?

It is not clear whether we can stop the process itself. But we have to ask ourselves whether we want it to happen haphazardly and have a repeat of the Middle East power shift and related human suffering, or whether we want it to occur within a global framework that would provide for sanity, prevent hawkish cartelization, and ensure continued supply to all countries.

The long-term self-interest of all countries lies in starting the conservation of resources immediately and in supporting international efforts in the same direction. Many developed countries will be especially vulnerable to future crises because their predominant lifestyles result in the use and waste of a lot of resources and they have already exhausted much of their own local supplies. Does the US want another series of wars as one metal after another runs scarce?

Countries will want to cut back production and use of metals by a fair amount in the medium term. Significant changes are needed. The reality of non-renewable resources is one of impending crises that can be softened and delayed with the proper global framework.

### Rate of Implementation

Two of the issues regarding the conservation of non-renewable resources are how far and at what rate we should cut back their extraction. No single person can determine that. In the long term, we would be looking at extended sustainability levels. In the immediacy, the answer will likely come as a result of a political and social process. To help in this, I would like to discuss a couple of considerations relevant to these issues.

The potential job losses resulting from the implementation of the tax shift is a legitimate social issue. Large-scale unemployment is not expected to result from that transition as the total level of taxation would not increase and the money supply would not change. Overall, the same amount of consumption and investment would be around, but it would shift sectors, for example, from oil to renewable energy. Total employment should therefore remain relatively stable, but work would not transfer directly from one sector to another. If you lose your job as an oil worker, you may not get one in the new wind turbine industry or get in at the same pay rate as you had. So, just how fast should we go?

Attrition has been used in the past as a worker-friendly approach to change and is usually one of the better alternatives whenever possible. Rather than laying off people, companies do not hire new employees as older ones retire or quit for one reason or another. Alternatively, they offer bonuses to workers willing to take early retirement. It is undeniably the best possible option when companies can afford it.

However, neither system is perfect. When a business closes down, there is no money for early retirement and everybody gets laid off, old and young. Even when planned, a closure by attrition or early retirement would leave some casualties. How long would it take to fully implement the ETS if we did it by attrition?

New entrants into the workforce may average about 20 years of age. Let's assume that most will retire at age 60. That leaves an average work span of about 40 years. Add another 10 years as a measure of security. As such, someone starting work today would technically retire in 50 years.

Consequently, at the best possible speed, we would need about 50 years to bring down production amounts of non-renewable resources from 100% (current levels) to 0% if we were to totally stop the extraction of minerals. Obviously, that is not what we are trying to do. If we were to bring them down to 50% of current levels in the medium term, it would take us about 25 years through attrition.

The above is not to suggest that we should reach that specific target in that span of time, but it is a benchmark that can help us see more clearly into the future and know more what to expect from a softer ETS implementation. We may choose to go faster for environmental expediency and especially at the beginning. Ultimately, the socio-political process will determine the exact speed of implementation of the ETS.

If half of one attrition span (25 years) is deemed both a reasonable and responsible time line to reach medium-term resource conservation targets, by 2033 we will have cut by 50% our consumption of non-renewable resources. Another half-span would take us to 25% of current consumption levels by 2058.

What targets countries will chose for the very long term is uncertain at this point. What we know for sure right now is that we should start early and speed up the political process. Our current rate of use is highly damaging to the environment and severely depletes resources for future generations. The earlier we start, the better.

### *Environmental Taxation Management Concerns*

Environmental initiatives can pose a number of problems if not designed appropriately. Alain Verbeke (Solvay Business School, Free University Brussels, Belgium) and Chris Coeck (Faculty of Applied Economics, University Centre of Antwerp, Belgium) are concerned that poorly managed environmental levies may result in a backlash in the business community, a decrease of their support for environmental taxation, and, in their words, the general impression that environmental taxes have become "arbitrary measures to stabilize government income" (Verbeke and Coeck, 1997, p. 510). In other words, politicians could abuse the system to raise taxes or make up revenue shortfalls.

Another concern raised in literature with respect to taxation is its dynamism. Verbeke and Coeck (1997) warn that using taxation as a source of income for governments or as funding for environmental programs may not yield the intended benefits for a number of reasons. Green taxation revenues tend to be dynamic and may not provide as stable a source of government income as desired. Let us look at an example of this.

Suppose that a government implements a petroleum tax that is expected to generate a billion dollars in revenue. The resulting increase in gasoline prices would generate positive environmental behaviors as expected—for example, the purchase of more energy-efficient cars, increased use of public transportation, a switch to alternative energies, etc.

Because this would result in a decrease in oil consumption, the tax would bring in less than the expected revenue target. Governments would then have to increase it further to generate the originally desired one billion dollars. This action would presumably lead to lower consumption yet. The environmental tax would have to be raised again. Dynamism would create some kind of *treadmill taxation effect*, which would eventually create a number of problems. Let's address these issues.

Firstly, the ETS is specifically designed to be revenue neutral. The proposed system is a tax shift and not additional taxation. As such, it would not be used to arbitrarily increase government revenue or make up budget shortfalls. Revenue neutrality is a large part of its political viability. Open disclosure and a high level of transparency would easily prevent abuse of the system.

Secondly, although current green taxation is often used to fund environmental programs (for example, taxing contaminants and using the proceeds to fund water improvement programs), all environmental levies from the ETS would be directed to general revenue accounts as is the case with current income and retail taxes. Government spending on social programs and services would not change. This would promote transparency and prevent the arbitrary financing of certain programs over others.

Thirdly, the treadmill-taxation effect Verbeke and Coeck describe is a very legitimate concern. Continuously increasing levels of environmental taxation could seriously undermine a country's international competitiveness. However, the dynamism of environmental levies is a very good thing. It would spur us on to continue year after year to improve resource conservation and environmental protection. We want the economy to become more and more environmental over time. Dynamism is the chisel which would make society and the planet greener and greener year after year.

The implementation of the ETS would be progressive. This implies that, in most cases, initial taxation levels would need to be further increased in order to reach long-term target levels. There would be plenty of room for raising tax levels to compensate for the effect of dynamism.

In the long run, revenues will stabilize on their own. In addition, once desirable levels of conservation and protection are achieved, or international competitiveness ceilings are reached, we can always fall back on single-purpose levies. That is, the dynamic effect of the ETS is perfectly and easily stoppable.

Dynamism is exactly what we need in order to continuously improve on conservation and the environment. We just have to remain aware that it will need to be managed. Note that competitiveness ceilings will only occur if the ETS is not implemented internationally. If worldwide or trading block agreements are reached on environmental taxation—which is likely to happen—they will be much less of an issue.

Under the current system, government revenue varies from year to year depending on profits, personal earnings, unemployment rates, national retail sales, etc. It decreases in times of recession and increases in strong periods of growth. Varying revenue is not a new issue for governments; gaps in income have historically been made up by temporarily borrowing or through budget cuts. As such, there is already a significant amount of variation of government revenues from year to year. The ETS would only continue that, not change it.

### National and International Issues

International competitiveness is probably the most difficult issue concerning primary resource taxation. Although most countries could readily initiate and gradually put in place an ETS strategy, few would be able to implement it at once on a full-scale because of the effect on their international competitiveness.

Environmental taxation is not unique in this respect. Minimum wage levels, interest rates, productivity, social programs, and a number of other factors also affect international competitiveness. The ETS would be just one of a range of variables already affecting international competitiveness.

Any resource tax would by necessity be collected prior to export, hence, make a country's goods more expensive and less competitive in foreign markets. Although exceptions could be made, a rebate on exports would not be a realistic solution as a resource tax would become diffused in finished products.

For example, a bar of steel may see a full increase in price from a given levy. However, a product with a 30% steel content in terms of value would see only a partial increase in price from the resource tax. For that reason, it would be far too complex to try to estimate the percentage of non-renewable resources in every item to be exported and compensate for the levy with a rebate.

Individual countries would therefore be limited in their ability to tax resources past a certain level as it would overly decrease their international competitiveness. However, initial implementation with lower tax rates would be doable for most countries and yield substantial environmental benefits. In addition, it would mean that fewer non-renewable resources would be exported, hence, that they would be saved for future generations. We do not want to conserve and tax resources in home markets but export them cheaply overseas to be wasted. Of course, countries could choose to not fully tax the export of raw materials in order to lower the impact on their market shares.

The ETS would also affect primary resources more strongly than final consumer goods. This would prompt countries to transform them more extensively into finished products and develop their local manufacturing industry—which would create jobs—rather than export them as raw materials.

Keep in mind that taxation levels would be political decisions and that an ETS implementation could be done slowly and not involve disruptions to major industries if desired. To soften the impact of resource levies on certain industries, governments would have the option of initially imposing import taxes on big ticket items so as to maintain local manufacturers' competitive advantage within a country. As more and more countries get involved in conservation programs, import taxes could be progressively phased out.

The long-term and most efficient solution to competitiveness differentials is the negotiation of international accords. Under such agreements, countries would see their taxation levels increase at the same agreed upon rate.

Because of its scalability, the ETS is readily implementable on a national basis in most areas of the world. Initially, it may need to involve a combination of resource and import taxes. Although limited by international competitiveness, resource taxation would yield very significant conservation and environmental benefits nationally.

The first countries to implement the ETS would also be the first to shift their economies towards cutting-edge sectors of the future (metal substitutes, conservational techniques and designs, alternative energies, etc.) and would therefore gain a huge competitive advantage in wide-open emerging markets. International agreements would remedy to competitiveness problems between countries and make it possible to shift into high gear the global conservation of resources and preservation of the environment.

### *Burden of Change*

Throughout history, all too often the cost of social advancement, the burden of change, has been borne by the few, those who lose their means of living as a result of economic events or industrial and social shifts. To a large extent, this is a choice. If society as a whole benefits from certain changes, why should a few be responsible for the costs associated with them? Why should we not compensate them for these?

Clearly, some of the costs of change cannot be avoided, but we, at least, can mitigate them. It is not only an issue of social justice; the absence of compensation is often counter-productive. The higher the unmitigated costs, the more reluctance to change is created in a society. In a constant race for economic positioning, a nation which resists change loses its edge and quickly falls behind. Tomorrow's innovative societies are those which will be best able to quickly adapt to new and changing environments. Removing blocks to change is becoming an increasingly important factor in this rapidly evolving world.

Unemployment in specific sectors or for specific people is often a direct cost of change. Improving the ways we deal with this problem would ensure that certain individuals do not unduly bear the burden for changes that benefit an entire society. In fact, it is the responsibility of a fair society to make certain that they do not. It would also help remove resistance to change, making a country much more innovative and successful in a competitive world environment.

## Implementation Scenarios

In the next few sections, I would like to take a closer look at theoretical scenarios of implementation for the ETS. This will help us visualize the new landscape and demystify certain issues. The first one is a sudden and drastic script that will serve to dispel some of the myths that may arise as a result of the work of lobbies opposing environmental taxation. Of course, I do not suggest that any country opt for such a scenario, but stretching ideas and using extreme renditions of a plan are ways to test their soundness and feasibility.

The second one is a blueprint for progressive national implementation that should familiarize us with the ETS's green economic environment and the various implications it may have for our lifestyles and our children.

All monetary figures in this chapter will be expressed in Canadian dollars (on par with the U.S. dollar as I write). As well, for the time being, we will set aside issues of international competition and focus strictly on national scenarios.

### D-Day Implementation

D-day implementation refers to a drastic scenario which involves high initial tax levels. It would occur on a national basis and assumes no international trade issues. Suppose that the Canadian government decides to implement an environmental taxation system overnight. The federal retail sales tax on goods and services (GST) as well as their provincial equivalent would be eliminated. That would automatically lower the price of goods and services by about 12%.

The non-progressive (flat) bracket of income tax would be dropped, resulting in a rebate of about $10,000 for the average worker in Canada. That is, there would be no more tax on the first $30,000 of net personal income. Someone having a gross revenue of about $40,000 would have about $30,000 net after basic-personal and other deductions (about $10,000) and, therefore, would not pay any income tax. As a result, an average worker's monthly take-home pay would be instantly about $800 more (the $10,000 rebate divided by 12).

The new ETS levies implemented by the government would include: a 100% tax on non-renewable raw materials (effectively doubling the price of steel, copper, nickel, etc.); a flexible tax added to the cost of oil to maintain its price around $150 a barrel (or any other price determined appropriate politically or otherwise—which could be more or less) and promote the development of the renewable energy sector; and a 100% tax on key contaminants and pollutants.

Supported by regulations, standardization would also be a government priority for resource conservation. Reusable containers would be tax exempt. Those that cannot be reused but are recyclable or made of renewable materials would see a $1.00 levy. Those that are not reusable, not recyclable, and not made of renewable materials would have a $2.00 packaging surcharge added to their price.

After a few months, the resource tax on metals would have filtered up. The cost of everything that contained significant amounts of them would have gone up. On average, the items with high metal contents would have seen a 20% to 30% increase in price while the costs of those with moderate amounts would have risen by 10% to 20%. Items with low metal contents would have barely seen an increase. The impact of the ETS would decrease with the amount of transformation raw materials would see and their total final value. For example, many luxury items would be less affected as a larger proportion of their costs comes from labor.

Household cleaning and other chemical-based products would have almost doubled in price because of the contaminant tax. At the end of the year, the taxpayer who had been $10,000 richer because of the income tax reduction would have spent $10,000 more on purchases. The government would have given from one hand and taken back with the other! That is exactly the idea behind revenue-neutral taxes.

The purpose of the ETS is not to raise more money for the environment or the government. It is to change the economic incentive structure so as to naturally deter unenvironmental practices and promote green consumer behaviors. On average, people would break even at the end of the year, but consumption patterns would change for the better. As unenvironmental goods would have become more expensive, people would have moved away from them. As green alternatives would have become cheaper, consumption would have naturally shifted towards them.

The ETS would have created a green economic environment in which one would be rewarded for doing the right thing (buying green) and punished for purchasing products that are not environmental. All of this would have happened without additional costs to the taxpayer.

The marketplace would have also significantly changed for businesses. Unenvironmental products would have become hard to distribute and less and less profitable. Green goods, which had been difficult to sell because of their higher prices, would have gained a competitive edge and seen their markets expand. Many products would have been redesigned with less or no metal as the manufacturing input had become more costly. Automobiles would have shifted down to smaller fuel-efficient and conservational sizes and would use substitutes to metals whenever possible. Green industry sectors would be on their way to becoming highly profitable.

Although not desirable, a drastic overnight implementation of the ETS would not have resulted in economic chaos. Some things would have become more expensive, but spending cash would have increased in proportion as a result of people having to pay less income and retail taxes. As such, total consumption (and money supply) would have remained the same. People would have been spending differently, not less. Their way of life would have changed but not overly dramatically. Their standard of living would be comparable to earlier on.

Under the ETS, we would see the world around us become greener and greener year after year. Less metal would be part of our lives. Contaminants and toxic compounds in products would decrease. Fossil fuel use would go down, less garbage would be produced, the environment would begin cleaning itself up, etc. We would see an explosion of alternative energy technologies appear and a battery of green products hit the market shelves, most of them cheaper than their alternatives.

## Habit Shifts

Some people would likely want to take maximum advantage of the new green economic environment provided by the ETS and change their habits. For example, they would consume a lot more green products and keep their cars longer or get much smaller and more fuel-efficient ones. Their new environmental habits would lead to significant savings. In the example above, their actual expenses may only increase by $7,000, leaving them with a $3,000 bonus from the $10,000 tax break they received on income.

Others would go with the flow and change their habits at about the same rate as most people. They would spend $10,000 more, the same amount they got back in income tax reduction, and come out even. However, their habit changes would be for the better. A lot of resources would be conserved and contamination decreased despite no change in their financial situation.

Some people simply do not care about the environment or can afford not to. They would continue to change cars as often as before, fill up landfill sites, and buy products in non-renewable packaging. They would find life more expensive, spending more than before on consumer goods.

Overall, the ETS would be neutral, but some could benefit and others lose. The difference is, under this system those who do the right thing for others and future generations are the ones who will win out.

### *Employment Shifts*

Assuming no international competitiveness issues, an implementation of the ETS would not cause an overall or widespread loss of jobs as spending and the money supply would not change. Unenvironmental sectors would see employment decrease, but green industries would have a corresponding increase. Job creation in these would make up much of the losses from the non-green sectors primarily because the ETS is revenue neutral.

The $10,000 extra paid in environmental levies would be received back in income tax and retail sales rebates. The money not used to purchase traditional non-green goods would now be spent in other sectors of the economy.

Let's illustrate all of this. Assume that, after the implementation of the environmental taxation system, someone buys a brand new car at a price of $25,000—which would include about $5,000 worth of resource tax built into the cost of raw materials. As the ETS is revenue neutral, the owner would get back in income and retail tax rebates the $5,000 originally paid. Of course, this would not happen within a single year because a car is a big-ticket item. However, eventually taxes would even out as people do not usually buy vehicles every year, and revenue neutrality should be achieved.

Those who already had an automobile and decided to keep it longer as a result of the ETS, would use some of their tax rebate for parts and repairs as their car gets older. The rest would be spent on other goods in different industries.

Those who did not already have a vehicle would have to resort to public transportation, which is another way dollars would be expended. All of that spending would create jobs in different sectors.

Similarly, higher oil prices would decrease demand for gasoline but increase the consumption of renewable energies. Jobs would decrease in the former—or stabilize as the demand for energy continues to grow—and increase in the latter. The bottom line would be a shift of employment in the economy, not a loss.

This is the process and incentive structure that would progressively and relatively painlessly turn an economy green.

### Competitiveness and Efficiency

The ETS could make a country less competitive internationally in some sectors. A global agreement in this respect would decrease or eliminate competitiveness issues between countries and serve to expand internationally the shift to green economics. It would also permit a fuller implementation of the system at the national level.

Competitiveness is often confused with efficiency. An isolated implementation of the ETS in a country would probably make it more efficient although less competitive worldwide. Resource taxation would undeniably make it more difficult for some goods to compete internationally as levies would be a built-in cost.

However, the products themselves may not have been built less efficiently. It is just that we would collect taxes from them rather than from the income of workers. Fixing a car is more efficient than exca-vating new resources, processing them, building a new vehicle from scratch, adding the old one to landfills, and polluting throughout the process.

In some cases, the new system would force us to use substitutes that are more expensive than their destructive alternatives. This could make goods pricier, but the difference in production costs would not be lost. In using more expensive alternatives, we would save some of the cheaper resources for future generations—including our own chil-dren and grandchildren. Wealth would be shifted to them, not lost. The ETS would make our societies more efficient in the present and preserve resources and heritage for our children and grandchildren.

In many cases, the ETS would lead to the development of more economical alternatives. For example, recent legislation outlawing chlorofluorocarbons (CFCs)—a major culprit in ozone depletion—led to the development of substitutes that are actually cheaper (Jackson, 2000, p. 390).

Many alternative sources of energy may become less expensive than oil once an infrastructure is in place and they are mass produced. Also, the real costs of fossil fuels are actually quite different from what you pay at the pump. If you factor in environmental damage and the trillion-dollar war premiums to maintain supply lines, the picture is very different. These costs are real; you pay for them indirectly.

The ETS would decrease competitiveness to some degree, at least until international agreements are reached. However, its net effect on an economy would be positive: increased efficiency especially in reuse markets, conservation of cheaper supplies of resources for the future, the development of less expensive substitutes, as well as an overwhelming amount of environmental benefits.

### The Forest Industry

Just as some sectors in the economy would decrease in size and importance under an ETS system, others, such as the forest industry, would thrive. The demand for its products would increase. That would mean growth and more jobs. Government regulations would need to be put in place to ensure proper management of the industry.

For example, clear-cutting practices would have to be phased out and replaced by better management options. Replanting would have to increase to not only renew the resource but also expand it to meet increased demand. Forest preservation for current and future generations would have to become a priority.

### What Needs to Happen

Let's look back at the initial planning exercise we did to determine what should happen under ideal circumstances. The ETS tax on non-renewable resources would reduce mineral extraction, which is what we said needed to happen if resources were to be conserved.

The ETS would result in higher costs for products with high metal contents or produced with toxic chemicals. This would shift consumption to greener alternatives, as needs to happen. Mineral exports would be less competitive and decrease as a result. This would further help us conserve non-renewable resources for future generations.

Higher prices for new cars would force us to buy second-hand and keep vehicles longer, hence, conserve non-renewable resources. The use of public transportation would increase, and the automobile industry would shift to building more conservational and environmental vehicles. All of these things also need to take place for reasons already explained.

The fossil fuel strategy would force gasoline consumption down, reducing greenhouse gases. High prices would promote a shift to alternative energies. Again, these two things are what we are trying to achieve.

The ETS strategy proposed in this book would deliver exactly what would need to happen under ideal circumstances. Of course, some job and economic displacement would be inevitable no matter how we decide to address the environmental problems on hand. A better unemployment insurance system would help us make the transition and result in a society being more competitive in the long term.

The ETS would make the changes that need to happen much less painful than the other options we have. The above example was one of drastic implementation, but countries will likely begin slowly, making the transition relatively easy.

We have the means to bring about large-scale environmental change; the only question remaining is, do we have the will? The short answer is, we do not really have a choice. Problems are already bad and will only get worse with further delays. The ETS will mean changes in lifestyles which, once used to, are likely to be not very significant in terms of personal discomfort. However, they will mean a lot for conservation and the environment.

We can act now; the ETS gives us the means to do what needs to be done.

# 10. A Blueprint for National ETS Implementation

In this chapter, I will look at gradual steps for the implementation of an Environmental Taxation System at the national level.

## First Step: Laying Out the Foundation

The ETS is scalable. This means that the initial tax rate on raw mineral resources could be 100%—a doubling of current prices—but it could also be 30% or less as necessary. As such, the environmental taxation system proposed here can easily be phased in slowly and progressively, making it implementable without an excessive amount of planning. In the beginning, it would likely be difficult to estimate the most appropriate level for each tax. However, because the system is scalable, it would not be necessary to do so.

The new incentive structure could be established initially with lower rates. This would familiarize the public with the implications of the ETS and the new economic environment. It would allow governments to see how environmental taxes behave and interact with each other. Most importantly, it would make it possible for countries to begin creating the green economic environment without delay. Of course, as environmental taxes are collected, the retail sales levies and the non-progressive share of income tax would be reduced in amounts equal to the increases having resulted from the implementation of the

ETS.

What is important is to lay down the foundation as early as possible to start preserving the environment and conserving resources. Initiating the implementation early would have several advantages: it would give us more time, allowing for a softer implementation, and dispel any hopes that a government commitment to the new future is only lip service. It would be a firm signal to the business sector that times are definitely changing and that companies should start planning for a green economic environment.

## Second Step: Short-Term Levels of Taxation

The second stage is intermediate in nature and would seek to establish a realistic national implementation. Countries would fully commit themselves to the new system and begin looking at setting appropriate long-term ETS levels.

Designated rates of taxation should be high enough to achieve significant amounts of resource conservation and pollution reduction but not so high as to overly affect a country's international competitiveness.

## Third Step: Progressive Implementation

This step would lead to the establishment of optimal national levels of taxation for the different categories of products and resources involved. It would essentially be a fine tuning of the tax rates developed under the second step. The resource and other ETS levies would be gradually increased to the new targets.

## Fourth Step: International Agreements

As national implementation would eventually reach limits, the fourth step would be the development of international agreements. These would support fuller national implementation as well as take resource conservation and the preservation of the environment worldwide.

## The ETS Diffusion Effect

The ETS would be charged on raw materials and other products and diffuse itself as it moves up into finished products. Even fairly high initial tax levels—if this is what a country chooses to do—would not result in excessive price hikes for consumer goods.

## Tax Diffusion Through an Economy

The ETS would spread in different patterns throughout an economy. At the one end, items like machinery, machine tools, and automobiles would probably be hardest hit by resource taxation as they contain large amounts of metal. A 100% raw materials tax would not automatically mean a doubling of their price. The parts and components of machinery and vehicles are not simply metal. They are technology, labor, and materials.

A kilogram of steel and a tractor part weighing one kilo are not the same thing. The part is metal that has been smelted, cut, or shaped into a specific design. In those processes, value is added to raw materials. For example, the part weighing a kilo may sell at double the price of its weight in steel. That tax would therefore apply to only half of the final price. Let us look at an example based on the drastic implementation scenario discussed earlier.

Suppose that a tractor made of 90% metal had a value of $100,000, with the cost of its metal parts being $90,000 and that of its non-metallic components, $10,000. Firstly, the tractor would not double in price from a 100% tax on raw materials as only its metal parts would be affected by the tax; the $10,000 worth of non-metallic components would not go up in price.

Secondly, the tax would affect only the raw materials, not their fabrication. Suppose that raw materials cost $40,000 and their manufacturing into parts, $50,000. A 100% tax on non-renewable resources would double the cost of metals from $40,000 to $80,000. The costs of manufacturing parts ($50,000) and of non-metallic components ($10,000) would themselves not be affected. The final cost of the tractor would have increased by $40,000 to $140,000, which is a 40% rise in price.

At the other end of the spectrum would be services, such as the legal, educational, and healthcare industries. As these do not generally involve the selling of material goods, their costs should essentially not go up. If anything, they should decrease as retail taxes are dropped. The same would be true for renewable resources such as wood products and foodstuff.

In between you have the vast range of consumer products that would see no or varying increases in price depending on the value of the non-renewable raw materials they contain as well as the amount of transformation these have gone through.

### A New Consumer Environment

Dropping the GST in Canada would make goods and services 5% cheaper. Doing the same thing with the provincial sales taxes would further reduce most prices by another 6% or 7%. That is a total of about 11% to 12%. As a result, under a 100% ETS scenario you may see the price for high-metal content goods increase by about 30% net. The cost of services and renewable resources would decrease by a few percentage points. Other items would range from no to moderate increases in price.

Many things would be cheaper, others, more expensive, but our total purchasing power would essentially remain the same as we would have more disposable cash from the rebate on income and retail taxes. This would mean a different way of life but not a lower standard of living. We would experience a different consumer environment which would lead us to buy fewer non-renewable goods and more renewable ones. Our buying patterns would change.

### The Packaging Scenario

In the packaging industry, a fairly drastic scenario of implementation would yield similar results. The ETS would, for example, bring in taxes on new containers. The environmental levy would optionally be backed up by regulations standardizing them. The products purchased by consumers would remain exactly the same. A fruit juice is the same regardless of what it is bottled in.

As all new containers would be taxed, companies would naturally and progressively shift to recyclable and reusable alternatives. The soft drink and beer industries in Canada and other countries used to function on that basis and still partly do so today when they reuse their own bottles. In terms of economic organization, this is essentially old technology.

What would be different under the ETS is how the recycling of empty containers would work.   It would be based on markets as opposed to the bureaucratic refundable-deposit system.  Empty bottles would be bought and sold as any other goods.  Of course, the old way of doing things would still remain an option in situations where it is desired or might remain the best alternative.

In some cases, packaging would make no difference, the alternatives being as good or better.  In others, it might do so.  For example, grocery bags may become heavier from the purchase of foods in reusable glass bottles and jars.   Lighter containers would remain, however, a choice for consumers.

One thing that would change in our lifestyles is that we would likely spend more time recycling, taking items over to depots or processors and selling them back for reuse or raw materials.  Those who would be too busy to do so, or would not want to, could forgo the cash and just leave them at the curbside to be picked up by recycling entrepreneurs.

### The Dynamism Issue

The chisel effect of the ETS in shaping a new society would be continuous.   Dynamism is a very positive and desirable factor in bringing about environmental change.  Metal content would decrease in many things as non-renewable resources would be substituted for by renewable or reusable materials.  Businesses would continue to seek to lower their costs by switching away from taxed inputs.  As the cost of toxic intermediate chemicals would go up, they would be increasingly replaced by more environmentally friendly alternatives. Processing methods would change and become greener.

A continuous incentive to do better is exactly what we want.  For the first time with respect to the environment, the issue would not be trying to get enough funding but keeping positive change from happening too fast.

# *11. Disposable Grandchildren: Packaging and Contaminants*

This chapter will cover in more details one of the most important sectors of the economy as it relates to the environment: packaging. Its products are pervasive in our societies. Furthermore, many are single use and, for that reason, extremely wasteful. As such, the packaging sector has a massive impact on the environment and is in need of major changes.

Currently, conservation includes various recycling and reuse programs. Renewable resources such as paper products are also the focus of recycling because of the cost of their disposal. The packaging industry is of direct relevance to resource conservation not only because of its use of non-renewable materials—such as steel and aluminum—but also because of its products' utilization of landfill space. Targeting the sector is crucial because of the sheer amount of waste it generates and because, if properly managed, it is one of the keys to resource conservation.

## The 20th Century Approach

How do we manage the packaging industry? Do we impose severe restrictions on it? Do we tax it or assess import tariffs? It is undeniably one of the most difficult environmental problems to handle. The way we currently manage packaging is nothing short of a crime against humanity.

Most consumer goods are used over a period of time, from a few seconds or minutes—for example, food products—to many years. Packaging serves once and is then discarded. How can we still be so widely using non-renewable resources for it? We consume tons of depletable materials—that will be desperately needed by future generations—for products that not only see virtually no use but also cost a lot of money to dispose of and will plague landfill sites for decades if not centuries!

The 20$^{th}$ century can make one claim: to have brought about the *disposable society*. By indulging in convenience, we are turning the world into a garbage dump, making our very grandchildren disposable. They will be left living in a highly polluted environment, and their bodies will reflect their surroundings.

Recycling and reuse programs will need to see significant shifts in approach and scale if they are to achieve effectiveness and produce reasonable results for the environment. Recycling is not the long-term solution to resource conservation. It is just a part of it. On the current scale, it only mitigates the problem although we may feel that a lot is achieved.

Plastic soft-drink bottles, for example, can be recycled to make t-shirts, carpeting, or pillow stuffing. However, most bottles do not get recycled in the first place and end up in landfill sites. Products made from recyclables eventually also end up in garbage dumps, only later. Steel containers can be resmelted, but there are transportation costs involved and it is an energy-intensive process. A lot of them do not get recycled in the first place, and even metal that has been given a second life eventually rusts away into the environment.

Recycling helps conservation, and efforts in this direction should continue. However, it only slows down depletion and delays the inevitable. For that reason, the only real long-term solution for the industry is to reduce, reuse, and shift to renewable resources such as cardboard.

## The Market Approach

Recycling programs often have to be funded by governments because there is generally not a market for used items at the actual cost of collection. That is, recyclables are most often resold by municipalities at less than it costs to pick them up. Taxpayers, therefore, have to make up the difference. Furthermore, the government bureaucracies that run the programs tend to be less efficient than their private sector counterparts.

Another approach to recycling is the refundable-deposit system in which, for example, a small levy is charged on bottles and cans and refunded when the empty containers are returned to vendors. Again, this is an inefficient approach. Hundreds of millions of deposits have to be collected and kept track of by retailers. Then, each has to be refunded. Net surpluses and deficits have to be accounted for and returned to or claimed from government agencies.

Environmental taxation would enable and support a new approach to conservation. The ETS levy would have the double effect of reducing our consumption of non-renewable resources and of creating natural markets for *reusables* and *recyclables* by increasing their value above their cost of collection. This would result in the elimination of the bureaucratic and inefficient deposit system.

Private companies would be able to buy and sell reusables and recyclables for profit. They would be collected and sold for cash at market prices without government intervention. Taxation would have the advantage of keeping the element of choice for both corporations and consumers. Certain useful but wasteful types of packaging would still be available, although for a higher price, as opposed to being regulated out. The approach would offer more options for consumers and is generally preferred by businesses. It provides them with more flexibility and time to adapt.

Natural markets would develop by themselves in the long term as non-renewable resource taxes are progressively increased and profitability levels are reached in those sectors, i.e. when the cost of recycled resources such as metals is significantly lower than that of newly extracted minerals. However, in the short term, there could be the need for a combination of approaches.

## The Short-Term Market Approach

Although resource taxation remains, in my opinion, the most efficient approach to environmental change, its initial levels would be limited by international competitiveness. As such, an interim strategy for packaging is likely to be needed in the beginning. Because it is a single-use item, packaging is extremely wasteful and should therefore be aggressively targeted from the start by any conservation approach.

The goal for the industry would be to shift to either reusable containers or 100% renewable, recyclable, and biodegradable resources such as cardboard. Non-renewable and non-biodegradable materials such as polystyrene fillers (better known under the trade name Styrofoam) would be taxed in order to foster their replacement by environmental alternatives such as cardboard frames, molded paper pulp, etc. Regulations could further be applied to inks, glues, tapes, chemicals, etc. to ensure that they are of only non-toxic and fully biodegradable types.

With respect to food-grade and other containers having a potential for reuse, the ETS would tax new items, making used ones or those made with recycled materials cheaper. Market forces would naturally act to shift industries towards them. Food-grade containers include the various plastic tubs and steel cans that edibles come in, as well as the aluminum cans and glass or plastic bottles used for liquids. The goal for them would be reuse as they are a large part of our daily production of garbage.

### Generalities

Note that under the ETS packaging strategy proposed here, environmental levies could be included in the retail price of goods or not. Some consumers prefer seeing on item tags the final price paid at the cash register. Others like better any environmental deposit or surcharge displayed separately. With the ETS, people would not return their empty containers for a refund. Instead, they would resell them to a local recycler for what these would be worth on the market at that time.

The role of governments would be to set taxes sufficiently high so that empty containers would be worth reselling instead of being thrown away. In some ways, the ETS approach would resemble the refundable-deposit system, except that it would not have its bureaucracy. One of the main differences between the two would be that the price of returns would not be fixed but vary from recycler to recycler and by locality.

Communities that are too small for recycling to be profitable or that do not have local recyclers could run programs themselves. They would set a price for recyclables high enough for people to have an incentive to return them. Losses could be made up by municipal, provincial, state, or federal governments, depending on agreements worked out to that effect.

## Double-Taxing Inputs

Under the general ETS scheme, raw materials (outputs) from producers would be taxed once. One way to solve the lack of initial incentive in the reuse and recycling industry would be to tax those outputs again when they are bought as inputs by packaging manufacturers.

For example, steel and glass would be taxed once with producers at the established ETS rate. Manufacturers that use these to make regular goods (tools, toys, tableware, etc.) would not pay a second levy. However, companies using the raw materials for the manufacturing of packaging would be taxed a second time as they are purchased as inputs.

This would make containers fabricated from new materials more expensive, which would increase the incentive for packaging manufacturers to move to used materials and create a market for recyclables.

Each type of material could be assessed for environmental friendliness or desirability. Criteria such as reusability, renewability, biodegradability, and toxicity could be used to set tax levels. For example, steel and aluminum would be assessed higher levies because they are depletable. Cardboard would be at the bottom of the scale because it is both renewable and biodegradable. As desired, the various types of plastics currently used in the packaging industry could be assessed individually and get different tax rates.

Imported containers could easily be taxed based on weight. Governments would simply have to require exporters to list the types of materials and net weights of packaging on labels and shipping documentation.

This would probably be the simplest approach to creating a market-based strategy for the packaging industry right from the beginning. In the longer term, the basic ETS scheme might be enough to support natural markets for used materials but could be enhanced with double-taxation to ensure high standards of recycling in an industry that is very wasteful and single use.

## Individual Taxation

A less desirable strategy would involve individual levies on types of items. This approach would be more difficult to implement because of the variety of packaging available. However, it may be preferred by some countries for one reason or another. Let us first look at the range of options that we currently have.

The very best environmental packaging at the moment is glass. It is natural, recyclable, and reusable. Under an individual taxation scheme, new containers would face levies to promote reuse and expand markets for recyclables. Used ones would remain unlevied.

Plain cardboard would be a very good second choice but can only be used for packaging dry goods. For liquids, an alternative to glass is waxed cardboard cartons, currently used in the packaging of milk and some drinks in many countries. These, however, are generally not reusable. They could be levied but at a low level because of their bio-biodegradability and the renewability of their materials.

Other alternatives such as regular plastic bags (those used for carrying groceries) and plastic-lined cardboard cartons could be levied at an intermediate level. They are less environmentally desirable but better than many other options.

Because they do not bio-degrade easily and are generally not made from renewable resources, plastic containers would be fully targeted by the ETS. So would tin and aluminum cans. However, items made from recycled materials would be taxed at a lower level in order to promote their use and the development of their markets.

### Examples of Individual Taxation

Most prepackaged foods and drinks would be targeted by the ETS. Double-taxation is, in my view, our best approach for the packaging sector and is so simple that no example is needed to illustrate it. Individual taxation would be much more complex. The following is an example of it and assumes that regulations standardizing container types and sizes have been implemented.

A typical scheme could look like this. Taxes would be charged either at the cash register or directly to manufacturers and imports at customs. The first level of levies, $1.00/item for small sizes, would be applied to new non-renewable, non-reusable, non-biodegradable containers. Larger or more expensive containers could be the object of a separate scheme or category. For the sake of simplicity, the following will deal only with smaller size items such as soft-drink bottles and tin cans for foods.

*Non-renewable* would refer to packaging made from resources that are limited in supply and depletable such as steel, tin, aluminum, etc. As most materials do eventually biodegrade, *non-biodegradable* would refer to those that do not readily break down in nature. The first level would essentially comprise containers that we would want to phase down or out for their unenvironmental or unconservational characteristics. This would shift producers and food processors—as well as consumers—away from them and towards better alternatives.

The $1.00 levy would not be a refundable deposit but a cost. It would be recouped through revenue-neutrality and by selling containers back to recyclers.

The second level of levies, $0.75/item, would be applied to composite containers such as plastic- or foil-lined cardboard cartons, for example, those used for packaging juices. This level would represent better alternatives. The levy would promote a switch to more environmentally desirable types of materials.

The third level of taxes, $0.40/item, would be charged on all new non-standard containers not already taxed above and made of reusable and recyclable materials. This would represent good alternatives such as glass containers. New and non-standard items would be taxed to promote a shift to standard ones, which would be more reusable, and to encourage reuse.

The fourth level would target new standard containers. A $0.30/item levy would promote their use over that of non-standard ones. The less fragmented markets are, the cheaper and more efficient processing for reuse would be because of economies of scale. This would result in lower costs to consumers, more successful processing industries, and increased resource conservation.

Fifth level packaging, used containers, would not be taxed. At present, this field is fairly limited. Beer is one industry in which bottles are washed, sterilized, and reused. However, this is undeniably the exception rather than the rule. Various foods could be packaged in reusable glass jars. Soft drinks could go back to being bottled as they used to.

The above could be a typical packaging tax scheme used to promote environmental and conservational behaviors in the sector. Of course, countries would define their own, and rates would be the result of a political process. As such, there should be no need to fear that extreme levels would be adopted.

As a general rule, the higher a tax, the more industries would shift away from undesirable products and towards more environmental alternatives. Countries could distinguish between types of plastics based on reprocessability. Reusability could also be graded and levied differentially based on the number of times a container can be refilled. Ultimately, each container type could even be graded individually based on environmental and conservational characteristics and desirability. So could raw materials under a double-taxation scheme.

Overall, the above system would mean that the choices that are better for the planet would be less expensive. Plastic soft-drink bottles would still show up on shelves; so would tin and aluminum cans. However, they would be more costly. That would provide for flexibility for both consumers—who would still be able to choose lighter packaging out of convenience—and producers, which may find it cheaper or more useful to use less environmentally desirable containers, or too expensive to convert away from them in the short term.

## International Issues

The ETS packaging component would be relatively simple to implement on a national basis. Once the taxes are in place, the market would take care of the rest. The question is, how do we ensure that local manufacturers and businesses are not put at an unfair advantage with the implementation of such a system?

Packaging does not represent a large percentage of the total value of the products we purchase. Furthermore, once markets are developed, used containers and those made from recycled materials may come out to be cheaper than even unlevied new ones. As such, there might not be a need to do anything.

However, imported packaging itself (empty boxes, bottles, etc.) would have an unfair competitive advantage. To ensure fairness, governments may choose to implement the levies that are applied to local packaging on all imports and optionally rebate them on exports. This could be done under either the double- or the individual-taxation scheme. Foreign manufacturers using recycled standard containers or countries having such programs could qualify for tax exemption through bilateral agreements.

A second option would be to apply a uniform but lower tax on all imports. This would offset some or most of the unfairness in competitiveness and keep things simpler.

A third option would be to charge levies at points of sales in stores. That would be much more bureaucratic, greatly multiplying the number of places from which governments would have to collect. However, it would have the benefit of treating both locally-made and imported products in the same way. The bureaucracy would shift from manufacturers to retailers. Such an approach could make it difficult to distinguish between new and reused containers.

When Canada shifted from collecting levies at the manufacturing level to doing it from retailers (under the Goods and Services Tax), it went through this. Stores had to reprogram their cash registers in order to collect the tax on some items and not on others. Many retailers currently charge a refundable deposit on certain types of bottle. Things could get complicated.

In the long run, the solution may lie with international agreements since common standards would greatly simplify things.

## Standardization

Taxes on packaging would shift markets to environmental alternatives. As already stated, glass is the best option as it is reusable. However, the fact that containers can be reused does not mean that they will be and will not go straight to the landfill. Many fruit juice and drink bottles are not reusable. Others are never recycled despite the refundable-deposit system. Discarded glass is just as unenvironmental as anything else.

The plethora of formats currently existing on the market makes it generally unprofitable to collect containers and process them for reuse. Unless special measures are taken to address the issue, we could still end up adding substantially to our garbage problem—not to speak of wasting a very useful resource—even if we have a sound recycling and reuse strategy. As such, countries serious about the environment may choose to support the ETS packaging component with regulations.

Currently, some soft-drink and most beer bottles are refilled several times before being recycled. In their case, it is achieved through a refundable-deposit system, which may not be the most efficient way to proceed. Countries may choose to not touch the industries that already reuse their own containers although most might want to move away from the deposit system to a market-based approach.

In any case, for the above companies as well as all others, regulations could be implemented to support broader reuse markets and higher rates of container return. The large variety of sizes and shapes currently in existence makes it difficult and often unprofitable to process them for reuse. The market for each type is very small, and trying to collect, process, and warehouse them would be costly given the limited volumes. Even in the beer industry, there are dozens of bottle designs although it does not appear to be so. Most are very similar but specific to companies, fragmenting the market and making it more expensive for businesses to reuse them.

If containers were regulated into a minimum number of categories and designs, reuse could become more profitable and attractive to businesses of all sizes. To promote larger markets, governments could standardize sizes and formats. For example, there would be one type of 5 ml, 10 ml, 50 ml, 125 ml, 250 ml, 350 ml, 500 ml, 1 liter, etc. jars and bottles. All caps and lids would have to be made of plastic rather than metal. Design specifications would be made available to both local and foreign container manufacturers so that overseas exporters who wish to qualify for a tax exemption could do so. A levy differential between non-standard and standard containers would be put into effect to promote the shift to the latter.

Standard containers would be slightly less expensive to produce than their non-standard equivalents as they would be manufactured and handled in larger runs. As most companies would purchase the same types of containers, their market size would increase and they could be more easily reused as the larger volumes would make it profitable to collect and resell them.

The processing, transporting, storing, and wholesaling of used standard bottles and jars would be much less costly as several companies within an area would share the same pool of containers, reducing inventory expenses and allowing their processing for reuse to be carried out locally. This would naturally promote a shift to them.

Furthermore, buying *standard* would mean buying green. As such, consumer preferences would likely shift towards this more environment-friendly alternative. Using standard containers would essentially be free publicity and a marketing advantage for most companies.

Under an individual-taxation scheme, governments could impose a higher levy on new standard containers—both local and foreign—to shift the market to used ones. Otherwise, we would be back to square one. Food processors and other companies would naturally gravitate towards the cheaper alternative, generating a demand for them, and essentially creating a reuse market.

Under a double-taxation scheme, the second levy would act to make new standard containers more expensive than used ones and promote a shift to the latter.

What would such an approach result in? Shelves in stores would look different. Groceries would be heavier to carry. Consumers would collect their standard used jars and bottles and sell them for cash to recycling firms that would process them for resale. Reuse would be achieved without the bureaucratic and inefficient need for deposits and refunds, or the involvement of governments. An enormous amount of resource conservation would result. Landfill utilization would significantly decrease. Intermediate chemical use would also decrease.

## Used Container Processing

The beer industry often reuses its own bottles. It manages to do so because its product is pervasive in society and major players dominate the market. Large numbers of containers go back and forth between brewers and consumers. That allows for economies of scale to take place. big companies have the capital to invest in bottle processing machinery (washers, sterilizers, etc.), but smaller ones often cannot afford it.

A case in point is that of a Manitoba micro-brewery which had to discard the bottles that were returned to them by the refundable-deposit system because of the prohibitive costs of the processing equipment. Many small towns would face a similar problem if packaging levies were implemented without standardization. In most cases, containers would have to be shipped out of town to larger centers for processing. The fragmentation of the market would add to costs, and the diversity of formats would make it unprofitable to collect many types of containers.

Packaging levies would reduce the variety of designs as companies shift to containers in lower tax brackets. The market would become less fragmented. As a general rule, the more formats, the larger the markets need to be for profitability. With fewer ones, smaller centers could process them for reuse.

## Supply and Demand in the Reuse Market

Since the ETS strategy does not involve a deposit system, the price paid to consumers for their empties could vary depending on supply and demand and by localities. As such, the market could not always be relied upon to set a price which would ensure that recycling and reuse do occur.

Larger communities should reach high levels of efficiency and be the most beneficial to consumers. Smaller and less competitive ones may have to be supported by regulation. Governments may need to establish a minimum amount paid per empty container. It would have to be high enough to ensure that they are returned and conservation strategies bear fruit.

That price would be determined by the socio-political process. A number of states in the US have bottle return systems. The average deposit charged is about $0.05. Although some programs are relatively successful, the rate of return of others can be as low as 60%. Canada's deposit fees range from $0.05 to $0.40. In my view, a minimum of $0.20 to $0.30 would be needed to provide enough incentive for people to return most of their empty containers and achieve acceptable recycling standards in today's context. In places where markets would not occur naturally and be profitable, municipalities could themselves take over recycling.

The ETS would yield very tangible benefits in the packaging sector. A lot more would be reused and recycled than currently is. Waste would become valuable. Companies would actually compete over your garbage. Resource depletion and landfill expenses would be greatly reduced, intermediate chemical usage from the manufacturing of new containers would decrease, and the cost of recycling programs to municipalities, eliminated in many cases.

As packaging is massively used everyday, governments should not hesitate to use high standards of reusability, renewability, and biodegradability in its respect. They should do so urgently. The industry is in dire need of a comprehensive policy ensuring that reuse is maximized and that wastage is minimized. The only issue here is convenience, and our inability to get our act together.

## ETS Management of Renewable Resources

Governments may also choose to tax some renewable resources to avoid over-exploitation. For example, new paper could be levied (at the producer level for collection efficiency) to reduce pressure on forests if proper management fails.

Recycled paper products would then become cheaper than new ones, promoting conservation and saving landfill space. A demand for used paper would be created. Market forces would kick in, leading companies into the business of collecting it and reselling it for profit to recyclers. Some of this is occurring at the moment but on a much smaller scale than it should.

The ETS could be pushed further into a full management strategy for renewable resources. To promote reforestation, governments could tax virgin lumber in order to shift the demand to timber coming from land that has been replanted. It could assess higher levies on lumber produced through clear cutting or other poor management practices (assuming those are not regulated out).

The same strategy could be applied to fisheries, with poorer management approaches penalized by taxation, or species with dwindling stocks assessed levies to increase their price and decrease demand.

The ETS would be a powerful market-based mechanism to manage renewable sectors of the economy. It would provide for them the same benefits that it offers as a conservation strategy for non-renewable resources: simplicity, flexibility, market friendliness, minimal bureaucracy, etc.

## The New Green Environment

Resource conservation would include a combination of taxes on non-renewable resources to decrease their use and create natural markets for recycling. A second set of levies targeted to non-renewable packaging would deter their use and reduce unnecessary wastage. Standardization and taxes on new containers would lay the foundation for a solid reuse industry and enable it without the need for the inefficient deposit-refund system.

Levies on specific renewable resources may also be used to promote conservation and prevent over-exploitation by giving a competitive advantage to recycled products. Most municipal recycling programs would be taken over by private entrepreneurs. The new market-based approach to resource conservation in the packaging sector would be much more efficient and effective than the current patchwork of taxpayer-funded programs.

Remember once again that all these taxes would fill governments' coffers and result in a reduction of income and sales levies to keep taxation revenue neutral. We would not pay more taxes overall. The ETS would slowly reshape the economy. Store shelves would take on an alien appearance at first. Some types of foods may look funny in glass jars. Bulk sections in stores would likely expand. Most soft drinks would return to being bottled in glass and be heavier to carry.

We would not have the diversity of shapes we are used to, but that would only mean saving resources and the environment. We would still see the familiar logos of food companies on glass jars. Our T-shirts may revert to being made out of natural, renewable, and environmental materials—cotton, silk, etc.—instead of recycled soft-drink bottles.

## Flexibility and Scalability of the Market Approach

Choice would remain a component of a market-based resource conservation strategy. Non-standard, plastic, and most other types of containers would still be legal but more expensive. A soft-drink company that does not care about the environment could continue to use plastic bottles, but its product would be pricier because of the tax on non-reusable containers. Likewise, consumers would be able to continue to buy unenvironmentally packaged goods, if for a higher price.

Choice adds flexibility to a system. As already stated, it is an option that is often preferred by businesses as it gives them time to adapt. The flexibility of a market-based strategy would make it more acceptable to everybody. The level of taxation as chosen by society would determine how much more one would have to pay for convenience. That is, the approach is fully scalable. An ETS strategy in the packaging industry would not mean radical changes, unless it is what people in a country want.

## Contaminants

The case for the reduction of pollutants has already been made many times in environmental literature. As such, the issue will not be rediscussed here in any detail. Contaminants would be a major target of an ETS strategy. Some issues—appropriate tax levels, international competitiveness, etc.—are very similar to those of non-renewable resources.

One of the differences with respect to contaminants is that they are more substitutable than metals. There are generally many alternatives available. Also, their contribution to the final cost of many products is often much smaller than that of metals. As such, their impact on the price of consumer goods would generally be less important. These are some of the considerations that would have to be examined closely in defining and determining policy.

Under the ETS, contaminants would be taxed to reduce their use and shift industries to cleaner substitutes. The determination of what should be levied and at what level is much beyond the scope of this book. Each chemical has its own properties and effects on the environment and would probably gain from being assessed separately.

Those are decisions that individual governments will have to make.

## Industrial and Domestic Contaminants

In the industry, undesirable chemicals and other compounds would be taxed at the producer level. So would the byproducts of the manufacturing process if they are problematic for our health and the environment.

Many of the pollutants plaguing the planet today are not industrial. They are not intermediate chemicals or byproducts of the industry. They are the very goods produced for us consumers. This comprises the various household cleaners, laundry detergents, and solvents we employ everyday. These chemicals would also be targeted by the ETS (taxed at the manufacturing level) as they are used massively, in millions of households around the world. They could easily see a doubling (or more) in price, but that would be up to us to decide.

## The Agricultural Sector

Ironically, the industry that puts food on our tables is also a very significant source of pollution. Modern intensive agriculture uses a variety of herbicides and pesticides that degrade our water systems and the environment. The ETS would and should target these to reduce their use and shift the industry and its R&D towards more organic alternatives and practices.

Close attention should also be paid to domestic herbicides and pesticides for the same reason. Some urban centers in North America have already made moves in that direction. These would fall under the contaminant strategy of the ETS and would be taxed at the manufacturing level.

A second source of pollution relating to modern agriculture is the widespread use of chemical fertilizers. These are responsible for the enormous growth in productivity that the industry saw in the last century but are also prime culprits in the degradation of our rivers and lakes. They would be targeted by the ETS to reduce their use and promote greener alternatives and practices in the industry.

An ETS strategy in the agricultural sector would raise the price of non-organic foods, shifting consumption to better and healthier organic alternatives. Revenue neutrality would mean that we would not be worse off financially.  We would only be eating fewer contaminants, antibiotics, and carcinogens while protecting agricultural land and the environment at the same time.

# 12. The Fossil Fuel Sector

Energy will be explored in conjunction with the automobile industry. Both have a massive impact on the planet and are undergoing significant changes.

Oil and other fossil fuels are currently vital to just about all countries around the world. Some are dependent on them to fill their energy needs. Others—like Canada, Mexico, Norway, and many countries in the Middle East—rely on them for income as net exporters.

Even within a country, world oil politics can have a tremendous impact. For example, most of the petroleum production in Canada currently comes from one part of the country, Alberta. Recent price hikes have meant billions of dollars in profits for that one province alone. A lot of that money came from the high prices charged for oil products to other Canadian provinces, some of which are significantly poorer than the already well-off Alberta.

The politics of fossil fuels can be complex and far-reaching. Nobody is immune to them.

## The Intermediate Phase

The oil industry is not about to disappear. It is just too vital to most countries. There is too much capital involved. The lobbies are too big and too influential to be displaced. There are also many mega-projects currently under development around the world.

From an environmental perspective, it would be desirable to phase out fossil fuels as quickly as possible. However, because of the reasons above, most changes will likely occur gradually. An intermediate stage for the industry would lead to a stabilization in the use of fossil fuels. During that phase, it is doubtful that a slowdown in production will occur given the continued growth of world population and the emergence of massive markets like China and India.

The wealth associated with this resource would not be lost, only partly shifted to future generations. The era of huge profits would be replaced by an intermediate phase of more moderate and less chaotic prices which would help stabilize the world and provide a steadier environment for doing business. Most countries are already committed to achieving either Kyoto greenhouse gas reduction targets or their own. This means that, one way or the other, fossil fuel consumption will be stabilized and reduced, whichever tool we choose to do it with.

The flexibility and scalability of the ETS would allow governments to define their own pace for the transition period and generally make it easier for the industry as a whole. The system is easily controllable and, again, would involve a political process that should ensure fairness and moderation in terms of target rates.

## The Market Approach to Renewable Energy

When the price of oil went up in the 1970s, many people and businesses bought into renewable energy technologies only to see their investment amount to nothing in the mid-1980s and 1990s when oil prices collapsed.

The ETS proposes to tax fossil fuels, such as petroleum and coal, for several purposes. Firstly, higher prices would reduce consumption, hence lower global warming pressures. Secondly, taxing fossil fuels at a level sufficiently high would make profitable the renewable energy industry and further the growth of a leading-edge economic sector.

The ETS would impose a variable levy on the price of a barrel of oil (either imported or locally produced) to raise it domestically to a set target, for example, $150.00. For a going market price of $120.00, the tax would add up to $30.00. It would be periodically readjusted to maintain the final price of oil at about $150.00 over time. This would create a predictable and stable environment that would not only foster the growth and development of alternative energy technologies but also prevent the crash of the new sector and the loss of precious green investment.

The ETS would be collected at the producer level. However, unlike other mineral levies, it would target domestic supplies and imports only. Exports would be exempt. This would raise the price of gasoline domestically and encourage conservation. But it would allow international market pricing mechanisms to continue to operate for petroleum, a resource vital to all countries.

Remember that the ETS is a tax shift. As the fuel levy is collected, income and sales taxes would be reduced in proportion.

## Benefits of an ETS Fossil Fuel Strategy

The ETS fossil fuel strategy would yield many benefits. The simple tax on petroleum would in one fell swoop conserve resources, reduce pollution and greenhouse gases, promote the development of the renewable energy sector, and decrease dependency on the Middle East. The total benefits are so large and far-reaching that we should immediately move ahead with the strategy. The ETS fossil fuel approach is not the path to the future; it is the eight-lane highway to it. Alone it would achieve the goals of the Kyoto Accord and yield multiple environmental benefits.

Currently, environmental strategies comprise most often a panoply of regulations and incentives such as grants and tax breaks, all of which require huge bureaucracies and a lot of monitoring. A tax on oil charged at the producer level where there are very few players would be much more efficient. With it, the law of supply and demand would on its own promote the development of the renewable energy sector without government intervention. This ETS strategy for fossil fuels would be efficient and produce a green growth sector that is stable and more competitive.

Fossil fuel taxes are never popular, but remember who pays for the grants and tax breaks used to fund environmental and renewable energy initiatives. We do. One way or the other, we pay. These expenses would disappear under an ETS system. In addition, with the ETS a lot of money that would otherwise be spent on monitoring and new bureaucracies would be saved. The end result would be a more competitive and stronger renewable energy sector that is fine tuned to market laws.

Other fossil fuels would also be taxed to reduce their use and decrease greenhouse gas emissions and other pollutants. The levies would prevent a shift from petroleum to other non-renewable types of energy which could be equally, if not more, damaging to the environment. Lower taxation could be applied to cleaner or less carbon-intensive technologies if these ever come through. For example, natural gas—a cleaner burning and over 50% less carbon-intensive fuel—could face lower taxes.

## Energy Future for Producers

Taxing petroleum is likely to be a very sensitive issue for producing countries. They tend to be very protective of their massive oil wealth. The ETS would not result in their giving away the revenue from these resources.

Internationally, the tax would not be imposed on exports, meaning that none of the profits from them would be lost. For example, Alberta, Canada, would continue to sell its oil overseas at international prices just as it does now. Nationally, world oil prices would also remain in effect. For example, Alberta would continue to sell its oil to Canadians at the international price just as it does now although there would be a tax on top of it.

Local producers and importers of oil would remain on a level playing field as the levy would be collected from both. As the ETS is revenue neutral, consumers would also come out even, paying lower income and retail taxes in exchange for the levy.

Specific producing regions would not be cheated through the new system either. For example, the levy that Albertans would pay to the federal government would not go to other provinces but would be received back as per the principle of revenue neutrality. The reverse would also be true; the levy that other provinces would pay the federal government would not go to Albertans but back to them.

What would change is that fossil fuel consumption would go down. It would undeniably be good for the environment everywhere, including producer regions. In the interim, oil prices would stabilize, which would benefit everybody, including the manufacturing sectors of exporting states and provinces. Oil not sold now will likely be sold tomorrow. As such, the wealth from producer regions would not be lost or given away but only shifted to their children.

## Ethanol-Blended Fuels

A 10% decrease in petroleum consumption in the transportation industry could be achieved fairly quickly with the addition of 10% ethanol to gasoline and diesel. These blends are cleaner burning and more carbon efficient than their pure forms. Current car motors and heavy-transportation diesel engines can burn these fuels without any modifications. E85 is an even better energy. It is a blend that contains 85% ethanol and only 15% gasoline but requires some engine modifications.

Cars can also run on 100% ethanol as is the case for four millions of them in Brazil. Furthermore, ethanol can use the existing gasoline distribution infrastructure. From that perspective, there are no legitimate reasons why these blend alternatives should not already be widely used.

### A New Set of Problems

Not everything is rosy about a biofuel strategy. The industry will have to be managed. At present, both the US and Canada subsidize their agricultural industries by the hundreds of millions of dollars every year because, among other things, of the low market price of some commodities. In Eastern Canada, potato crops were often destroyed to prevent an oversupply and a price collapse. Now, they could be turned into ethanol.

A biofuel strategy would be very positive in those sectors, and things have already begun to change. In the fall of 2005, France announced that it would turn part of its overly plentiful wine production into ethanol. Canadian and American farmers are already doing better from the rising demand for biofuels. The shift to renewable energy would reduce waste and losses and generally make agriculture much healthier financially. The new economic landscape would eliminate the need to subsidize it with taxpayers' money. It would also result in cleaner burning fuels, lower greenhouse gas emissions, and a reduced dependency on oil.

Unfortunately, not everything is good. Food prices are rising partly as a result of crops and land being diverted to the production of ethanol. Italians made headlines early on for complaining about a sharp rise in the price of pasta. Bakers in Quebec, Canada, are already seeing a doubling and tripling in the price of flour.

The United Nations is talking of a potential food crisis in developing countries as people can no longer afford the skyrocketing price of rice and world grain reserves are at historical lows. To some extent, it is already taking place. This is the result of a number of factors: rising oil prices, the production of biofuels from edible crops, and speculation.

## The New Economic Reality

Biofuels present a number of challenges. There is obviously the question of competition with food crops for arable land. But this does not have to occur. Biofuels can be generated from garbage, inedible plants grown in unused fields, and agricultural and industrial byproducts. Governments would therefore be able to intervene and exercise some control on the problem.

That would be important, especially in developing countries where many people can already barely afford food. The use of edible crops and good agricultural land for the production of biofuels could and should be restricted as necessary.

A second issue is the problem of low net efficiency: it takes a lot of energy (often from fossil fuels) to produce them. That should improve with technology, and things will sort themselves out through markets. Biofuel sources with low net efficiencies will be less profitable and, as a result, attract less investment or simply be abandoned. The most viable ones will emerge and grow.

Unlike biofuels, electricity generation from hydro, wind, tidal, geothermal, and solar sources is not expected to affect the price of food. It does not compete with it for land. Not all new alternatives would cause problems.

Biofuels and other sources of renewable energy will be an evolving landscape for some time. We should not panic as new problems and challenges emerge.

## *New Land for Agriculture*

Can new land be developed for the production of energy? Areas that are valuable to us for one reason or another (old-growth forests, wildlife habitats, etc.) should be protected, but could other types of land be converted and have a positive effect in terms of carbon reduction? A close look at the economics of land conversion seems to indicate that it could.

Large tracks of tropical forests are currently being cleared and planted with palm trees for the production of vegetal oil for biodiesel. Wetlands and grasslands are also being targeted for conversion to biofuel production. Unfortunately, these are already carbon sinks. As such, many argue that their clearing would add to global warming problems (Common, 2008, January 25).

The initial cutting of trees and brush would certainly contribute to the greenhouse effect. However, each harvest cycle itself would be carbon neutral as crops would be regrown, and the biofuel produced from them would decrease emissions from fossil fuels on an annual basis. As such, a reduction of greenhouse gas emissions would occur after a certain period of time and continue yearly after that. Of course, the degree of success of land-conversion initiatives would depend on a number of factors, among them, the various efficiencies of individual crops and production methods.

The development of new land for the production of biofuels could technically have a positive effect on global warming. However, in my view, we would be making the same mistake as we did before: treating the symptoms rather than the disease. Clearing strategies would lead to the progressive *denaturalization* of the planet by the replacement of forests, wetlands, etc. with crop fields. Is this really the answer, or is it the problem?

What put us into this mess in the first place is unbridled consumption and population growth. A permanent solution to environmental problems would need to address these issues.

The new economics of the environment will have to be closely monitored and regulated. The science will evolve as we go along, and governments will find out what works and what does not. Efficiencies will likely improve over time and with economies of scale. The new green environment would certainly create new challenges and call for changes in laws and regulations.

In the long run, food prices will probably trend higher not because of the biofuel strategy but because the cost of oil will keep increasing. This is just the new reality of shrinking supplies. The ETS would slow down that process by promoting alternative sources of energy, hence preserving oil reserves and slowing down the increase in price of the commodity. In addition, it would favor green goods and make them cheaper. And, that includes food. The ETS may just be what the world needs.

## A Market-Based Biofuel Strategy

Regulations that would force the addition of 10% ethanol to all gasoline were recently proposed in Canada. This type of strategy is certainly positive but could cause problems in that the oil industry would have to purchase quite suddenly a large amount of ethanol for which there is little supply at the moment. This could result in high costs for the fuel and chaotic gasoline prices.

The best approach could include regulations, especially initially, but should be based primarily on market forces. An high and stable oil price would make ethanol blends and biodiesel more competitive, which would promote a natural switch to these cleaner and more environmental options without creating a supply crisis. The higher the target price, the more competitive biofuels would become. As such, governments would be able to control the speed of the process.

A market approach would ensure a gradual and smoother transition by allowing bioenergy production to grow and respond to demand as opposed to using regulations and creating a supply crisis.

A biofuel strategy would create markets, sustain crop prices, and prevent the loss of investment and jobs. The growth of the bioenergy industry would ensure a thriving agricultural sector for the future and help redistribute wealth regionally. This is already happening. The ETS is exactly what many farmers and food producers need.

## Hydrogen Transportation: Panacea or Illusion?

Hydrogen has lately been the focus of much media attention. This potential energy of the future has especially captured the imagination of the public because its combustion or use in fuel cells to create electricity is essentially pollution free, water vapor being the only exhaust emission.

The main reason why hydrogen seems to have overtaken electricity —another perfectly emission-free energy—in the race for clean transportation is that it provides a longer driving range to vehicles. The best electrical storage technology (batteries) to date may satisfy urban commuting needs but remains impractical for longer range or heavy transportation.

There are a number of problems with hydrogen. Firstly, the gas is highly explosive. Safe storage technology is under development, but costs are and may remain an issue. A second concern is that it is only an intermediate fuel. There is no source of hydrogen per se. There are no vast pools of it underground like there are for oil. It needs to be produced from or with other sources of energy, for example, natural gas or electricity. An obvious issue with the former is that it is a fossil fuel, and producing hydrogen from it would release in the atmosphere large quantities of carbon dioxide, a greenhouse gas. As such, it would not be an alternative or a renewable energy.

Hydrogen can be produced with electricity through a process called electrolysis. However, each time you transform an energy into another, there is not only the cost of doing so but also a conversion loss. 100% of electrical power could yield only 70% of another energy. You are better off using it directly rather than converting it into something and then back into electricity in the fuel cell of a car. In the specific case of hydrogen, energy losses are currently substantial (Hydrogen Fuel Cells, 2004, October). Different avenues are being explored for its production and efficiencies are increasing, but it is unclear at this point in time whether or not these will be high enough to make hydrogen an energy of the future.

A third concern is the massive amount of capital investment a conversion to hydrogen fuel would require. Simple, well understood, and cheap infrastructure already exists for electricity: knowledge, distribution lines, motors, etc. That is not the case for hydrogen.

Capital expenditure for a hydrogen society would be high and probably fairly risky as a lot of government planning would be involved. It would require a huge commitment and a leap of faith into a future that may never happen.

A fourth concern is the old chicken or egg problem. Which came first? Do you build a huge and expensive distribution system first, hoping that people will switch to hydrogen transportation? Or do you build cars for which there is yet no fuel distribution system?

Breakthroughs may change the odds for hydrogen, but currently there are still too many questions left unanswered. The gas may only turn out to be a clean front for dirty fossil fuels. An educated guess at this time points to limited applications and niche markets for hydrogen.

A large-scale future for this fuel may come through if research leads to ways to convert the world's massive coal resources to the gas without carbon emissions or pollution. But again, even the large reserves of the fossil fuel will also eventually run out. One way or the other, a firm commitment to a hydrogen future may not be possible for another 10 to 20 years, until more research has been done.

## Conclusion

Compared to the renewable energy sector, the fossil fuel industry is highly concentrated. Its immediate and long-term interests are not necessarily the same as those of society as a whole. Global warming problems preclude our continuing down the fossil fuel road. Diversifying into decentralized renewable energies would spread wealth around and benefit future generations.

# 13. The Automobile Industry

## A Blueprint for a Renewable Energy Future

The Fourth Wave is not about a blind leap into the future. It is about large-scale but feasible change in the present. Hydrogen represents a vast field of research in itself, not to speak of the entire renewable energy sector. A comprehensive strategy could be the object of several volumes.

Doing a detailed study of renewable energies is beyond the scope of this book. Rather, I will try to identify major trends and highlight the issues and factors that may enable us to select policies for immediate implementation and develop a blueprint for the energy and transportation sectors of the near future.

Special attention will be paid to strategies that may benefit us on several counts as opposed to those furthering a single goal. Policies that may be able to bridge us to the medium-term will also be given preference. We do not want to develop an infrastructure that will become obsolete in 20 years when a final decision can be made on the hydrogen future. Decisions regarding a future for energy may ultimately be a matter of social choices as the various options open to us have different implications.

### The Hydrogen Future

At the moment, there are two major trends in the future of energy: hydrogen and biofuel-electricity. The hydrogen future is still some 10 to 20 years away. It is at an early stage of development and depends on new and untested technologies. It would be high risk as it would necessitate a huge investment in infrastructure and so far relies on fossil fuels for feedstock.

Hydrogen would see centralized and large-scale operations in both urban commuting and long-range transportation. The sector would likely be owned and dominated by a few large corporations as the technology and infrastructure require huge capital investments. While the wealthier parts of the world may be able to afford such a strategy, most countries may find it simply too expensive or not the most cost-efficient option for them.

### The Biofuel-Electricity Future

The biofuel-electricity future is actually a well-trodden path. It is based on well-established technology although research will continue to improve methods, production techniques, and hardware. Vegetable oils, from which biodiesel can be derived, have been produced for a long time. Ethanol—drinking alcohol—is actually something that has been around for millenniums. Its production is well understood and, at least in the case of Canada and the US, would be of prime benefit to the agricultural industry.

Current petroleum-based motors can already burn ethanol blends. Hybrids capable of using either gasoline, pure ethanol, or blends are currently available on the US market. Biofuels are carbon neutral. Their future would entail much less infrastructural spending than hydrogen as a lot of the technology is already here and can use existing distribution systems.

Brazil has already tested and proven the feasibility of a large-scale implementation of a biofuel strategy. Currently, the South American country has about four million cars running on ethanol produced from local sugar cane crops. The Brazil experience has greatly benefited its people by creating jobs, decreasing dependency on foreign oil, and improving the country's balance sheet.

Implementing a biofuel strategy would be relatively simple. Governments would set an oil price target that is high enough to make biofuels competitive. The market would take care of much of the rest. The strategy could be as gradual or as fast as we want and essentially without quotas or regulations. Moreover, it could be implemented nationally without international competitiveness issues, just like Brazil did. This is another eight-lane highway staring us in the face while we are still wondering what we should be doing.

## The Common Ground

A biofuel-electricity strategy would rely on ethanol and biodiesel for long-range transportation. Urban commuting would be based on electrical vehicles. Their shorter range (100-150 km) does not currently make them a realistic option for long-distance traveling, but their lack of emissions makes them perfect and as clean as hydrogen for commuting to work.

An electricity-based urban transportation strategy would greatly decrease the demand for petroleum, which would serve to keep its price at a moderate level. It would also reduce the need for biofuels, whose production competes for agricultural land and causes significant increases in the price of food.

Future technological directions are paramount in considering options for long-term strategies. Optimally, current choices should support future options whatever they may be. An interim strategy should yield an infrastructure that would be able to support both a hydrogen and a biofuel-electricity future. Options would vary depending on the wealth of individual countries and the availability of natural resources.

Currently, world energy is derived from a variety of sources: oil, natural gas, coal, hydro, etc. This is for good reasons: needs are so massive that concentrating on only one type would result in a supply shortage and skyrocketing prices. We have to continue to get energy from several different sources in order to fulfill all our needs and keep prices relatively stable.

In addition, different types are better suited to or more efficient for certain applications. For these reasons, we are likely to continue to draw energy from various sources in the future and should prepare for more than one option. In order to do this, one has to look for common grounds between the hydrogen and biofuel-electricity scenarios.

## The Convertible Electrical Vehicle

Hydrogen can be burnt in engines just like gasoline. Its combustion is clean, producing only $H_2O$, or pure water, as emissions. However, instead of being burnt, it can also be used in a fuel cell to produce electricity as power source for a motor. A fuel-cell car is actually an electrical vehicle powered by a hydrogen cell as opposed to a set of batteries. Fuel-cell and battery-powered cars could share the same architecture with the exception of one component, the energy module. Their motors and other components could be exactly the same.

Designing and developing convertible electrical vehicles may be the solution to the *chicken or egg* problem posed by a monolithic hydrogen strategy. New cars could be designed to use either batteries or fuel cells as source of energy. They could be converted from one to the other by the simple replacement of the power module. This would leave all the options open for the future, whatever scenario eventually comes through.

## A Second Common Ground: The Electrical Grid

The power grids would essentially be the only infrastructure needed to support electrical-vehicle transportation in urban settings. Batteries could be recharged at off-peak times (either at home or work) and prevent wastage.

The electrical grid is already a common infrastructure for a number of energy sources from which electricity is currently produced, for example, hydro, coal, or nuclear power plants. The same grid could also support many of the future's renewable energies, such as solar and wind power. If there is a significant shift from diesel and gasoline to electricity, the consumption of the latter would increase and drive prices up. New sources will be needed to meet the rising demand and keep the cost of electricity down.

As such, electrical networks would become central in increasing supply. The grid already provides support for renewable and decentralized energy production. With little additional investment, it could also support millions of micro-producers—for example, anyone purchasing a backyard wind turbine or solar panels with the intent of selling surplus energy into the grid. Currently, most power producers do not offer the option of buying back electricity, but that could easily change and it is starting to. With increased demand, micro-producers would become an important source of renewable energy supply. Wind turbines may just become an important part of both the rural and urban landscapes of the future.

The grid would become central to not only increasing the supply and delivery of renewable energies but also supporting hydrogen without the need for massive infrastructural investment. The clean fuel could be produced at home from grid electricity in small electrolysis machines. At the moment, efficiencies are not great, but that may change in the future.

Direct home production would remove middle-agents and retailers from the equation, meaning that it could favorably compete with commercial ventures. It could also take advantage of off-peak rates, which would also serve to lower costs. In addition, hydrogen produced from cleaner grid electricity would likely be a lot greener than the one extracted from fossil fuels. Of course, that would depend on how the various technologies involved develop. The future holds many promises in this respect.

The electrical grid is a common infrastructure that is already in place and that would offer the possibility of significantly increasing energy supplies and of doing so from renewable sources. Hydrogen would then have a green future rather than being produced from depletable and carbon-intensive fossil fuels.

## The Automobile Industry Under the ETS

A pragmatic ETS strategy in the automobile industry would firstly involve a tax on oil, as already stated, to stabilize its price at a higher level. The levy itself would have no bearing on whether we eventually move towards a hydrogen or a biofuel-electricity future. Secondly, it would involve the implementation of a blended-fuel strategy as it can be readily done without significant infrastructural changes and it also has a bearing on future energy directions.

A third step would be to select and implement strategies for the next decade or two, after which we should have the final word on the feasibility of a hydrogen future and would be able to firmly commit to a specific path. These strategies, while providing benefits now, would support all future energy directions.

There are two main sectors in transportation.  Each is qualitatively different in terms of needs, challenges, and markets.  Long-distance transportation, such as the trucking industry and holiday traveling, requires larger vehicles capable of going over long distances.  There is less common ground in future directions for this sector, at least for the moment.

Electrical vehicles are not currently appropriate for long-range and heavy freight and would not be suitable alternatives for these purposes at this point in time.  As such, long-range transportation would continue to be based on gasoline, diesel, and blends that include renewable fuels.

Urban commuting, the second main sector, is characterized by a multitude of smaller vehicles operating within densely populated areas.  These play an important role in urban air pollution and smog problems.  As such, everything should be done to make their emissions as clean and pollution free as possible.  The sector would be an ideal candidate for electrical or hydrogen-based vehicles.

### Transition in the Automobile Industry

Both the hydrogen and the biofuel-electricity future would find a common ground in a convertible electrical vehicle.  Governments and industry could work together in order to design a common architecture for modular electrical vehicles that could be powered by either batteries or fuel cells.  These would initially be operated with the former, making them highly suitable for urban commuting.  They would immediately provide cleaner urban environments and decrease fossil fuel consumption and greenhouse gas emissions.

Ten years from now, if and when hydrogen technology comes through, the same modular vehicles could be converted to fuel-cell power with the simple removal and replacement of the battery module.  These commuter cars would provide continued work in and a transition for the industry.

This would mean a cleaner environment now and for the next 10 years.  It would also create work for the automobile industry and provide a transition to both battery and fuel-cell technology.  This strategy would also conserve an enormous amount of non-renewable resources as an entire generation of vehicles would only have to be upgraded as opposed to being scrapped and added to landfills.

Convertibility between electricity and hydrogen for vehicles would also allow us to switch easily between different types of energy according to supply and cost. This would support a better and more stable price environment, provide a flexible and diversified energy strategy for the future, and perhaps prevent our being hostage to fossil fuels ever again.

### The Chicken or the Egg

From a business point of view, convertibility would provide the egg to the *chicken or egg* problem. Ten years from now, battery-operated but hydrogen-compatible vehicles would already be in wide use. Infrastructural investment would be less risky and could be provided by the private sector instead of the taxpayer and governments in high-risk ventures. To the market, it would be a simple investment decision that would depend on the price of hydrogen relative to its alternatives.

Decisions for the future of long-range transportation could be made later as new technologies develop and more information becomes available. In the meantime, the immediate implementation of a convertible electrical vehicle strategy for commuting would revolutionize the urban environment. Cities would quickly become much cleaner and quieter. Smog would be significantly reduced or may disappear altogether.

During the transition period, we would live in cities where single people would use mass transit or electrical vehicles to commute to work and rent a hybrid or gasoline car for holidays. Couples with two automobiles might keep a gas vehicle that one person would use to commute to work and that the family would drive on holidays. The second one would be an electrical commuter.

Appropriate electrical grid policies would have to be implemented alongside an electrical vehicle strategy in order to increase the supply of electricity and promote renewable sources of energy.

Despite the uncertainty of future developments and directions in the energy and transportation sectors, concrete steps that would reduce fossil fuel use can be taken immediately. The options we have offer both centralized and decentralized possibilities that may appeal to different countries.

# 14. Transportation

In this chapter, I will explore in more details future options relating to the automobile industry, a very important sector in the economy of many countries and one that will see plenty of changes.

In the short and the medium term, the ETS would increase the demand for more environmental and conservational generations of vehicles. In the long term, two trends would develop. Firstly, we would see part of the production shift to remanufacturing. Automobiles would be kept longer, fixed, and upgraded as opposed to being built from scratch and bought new. Greater standardization and increased modularity of architectures would make it easier for parts to be replaced and reconditioned.

Secondly, as a result of the taxation of non-renewable resources, a second trend would emerge: downsizing to single-passenger electrical vehicles or SPVs. These two avenues of development would spell a significant decrease in resource depletion and major changes in our cities and living environments.

## The Hybrid Question

What about hybrid vehicles, those having both a fuel engine and an electrical motor? They are often viewed as being the solution to all problems in the automobile industry of the future. Undeniably, they are a step forward. However, there are several issues in their respect. Early models really disappointed in terms of improved fuel consumption.

There has been much improvement in the technology since, but they may fall short of how far the transportation of the future needs to go. They will likely retain a place as family or holiday car but are not enough with respect to urban commuting. We need to do a lot better.

As well, hybrids use up more resources as they call for two engines instead of one and are heavier and more metal intensive as a result. This will probably limit their future.

In the long term, they will be able to carve themselves a share of the market if fuel efficiency increases significantly and the technology gets a lot smarter. That has already begun to happen. Some of the more promising developments lie with having smaller fuel engines and electrical motors run simultaneously and with technology able to recuperate the energy from breaking.

The future of the hybrid will largely depend on the industry's ability to develop its technology further.

## Impact and Directions

An environmental revolution would not signal the end of the automobile industry. For the last 20 years, there have been many calls for better public transportation in order to mitigate environmental problems.

While the approach can be effective in large cities, it is not the answer everywhere or to everything. The reality is that personal vehicles are here to stay and the automobile will remain pervasive in society. Public transit has to be improved, but more effective vehicles have to be designed for individual transportation. These are the two main avenues for the future.

## Public Transportation

Mass transit is an important alternative that offers multiple benefits to society. It reduces traffic and its related problems, for example, road congestion and air pollution. Most larger cities today would grind to a halt without it. Public transportation reduces the demand for and use of fossil fuels and promotes the conservation of non-renewable resources.

It prevents the use and purchase of millions of vehicles, and buses, trains, and subway cars are built to last much longer than regular automobiles. Because of their initial price tags, public transit vehicles also tend to be repaired and refurbished more extensively. That makes them highly conservational, several times more than current automobiles.

There are different mass transit models currently in effect in cities around the world. Some involve standard fares regardless of the distance traveled. Others are based on a concentric zoning system expanding away from city centers. Each has a certain amount of built-in inefficiency. For example, the standard-fare system charges the same price to people traveling short distances as it does to those transiting much longer ones. The zone model addresses this by setting fares generally based on the distance traveled from city centers, but does not reflect the intensity of travel routes.

In most cities, some transportation lines are heavily used while others are not. The buses, trains, or subway cars servicing the latter often run half empty or worse. That is not very good. Despite the longer distances involved, major routes can be far more efficient than shorter ones because vehicles are in full use. They generate more revenue, and the actual cost per person—and to the environment—can be much lower than what passengers are actually charged.

A third model better reflects actual costs and is also more environmental. The zones of the second option are modified into a more organic artery system. The efficiency pattern of public transportation systems is brachial just like a tree, with trunks and major branches being highly efficient and smaller ones being less so. The artery model would make many long-distance routes cheaper and encourage people living in suburbs to leave their cars at home and reduce pollution.

Instead of being concentric, transit zones would extend like fingers from the central business district along main arteries. Fares would be cheaper on primary routes as these would be more extensively used. They would increase on secondary and tertiary lines.

Public transportation is a clear avenue for the future and deserves government support. It has many limitations, for example, with respect to availability in suburbs and smaller towns, frequency of service, and practicality in certain situations. Government support in order to increase its use and benefits would mitigate many of its limitations; more people riding would mean more frequent service and route expansions.

As public transit will not satisfy all the transportation needs of the future, other ways of improving traffic, reducing pollution, and conserving resources have to be explored.

## Cycling and Car Pooling

Of course, one of the most obvious means of transportation is the bicycle. It is heavily used in a number of Asian countries. Unfortunately, it is increasingly being replaced by scooters, motorcycles, and automobiles—which are not as good environmentally.

Bicycles are not as popular in advanced countries although they provide a cheap, environment-friendly, and healthy alternative in terms of exercise. Many cities lack appropriate paths for them, making their use dangerous. The bicycle would do better under the ETS and would gain from being promoted by governments, be it in the form of paths, safety measures, or financial incentives.

Car pooling is a very good and growing environmental option, but it is not suitable in many situations and markets.

## Individual Transportation

Undeniably, the conservation of non-renewable resources would mean that automobiles would be kept longer and repaired more extensively. Many jobs would eventually shift from the new car industry to the parts and repair sector as well as pre-owned vehicle retailing. Refurbishing used automobiles with new interiors would be a growth industry in the medium term, assuming appropriate taxation levels.

Increasingly, cars would switch from being a disposable good with a seven or eight-year lifespan to being a semi-permanent vehicle that is fixed, upgraded, and refurbished for a couple of decades or longer. New cars would cost more, but their resale value would also be higher. Owners would maintain and upgrade them. The market would operate a little like real estate does now, where houses are renovated and the majority of properties bought and sold are not new.

Obviously, less metal would be used in designs. Vehicles would be smaller, lighter, and R&D would shift towards substitutes for metals: plastics, fiberglass, carbon fiber, composites, and biomaterials (for example BioFoam, a polyurethane foam made from plant seeds). The industry would focus on longer lasting and higher quality vehicles.

It would also move towards more easily reusable uniform chassis (where most of the metal in a car is located) and modular designs. Cars would still look different on the outside but would be built on similar basic architectures and with more standard components. These would make them more repairable as widely-used parts are less likely to go out of production.

Consumers would keep their cars much longer and likely buy used or refurbished vehicles. Eventually, companies would move into remanufacturing, where automobiles would be reconditioned anew and upgraded to the latest styles, as already expressed. For example, the entire interior could be redesigned. The exterior could be modified or if modular, changed outright. As time goes on and architectures become more standard, new cars could even be built on used frames.

Under the ETS, several generations of smarter vehicles would be developed. Consumption patterns and manufacturing techniques and practices would also change greatly.

## Conservational and Environmental Cars

Although electrical vehicles are not yet suitable for long-range transportation, the technology is essentially ready for the urban environment. Prototypes were built over 50 years ago. At this point in time, the best battery technology still leaves us wanting in terms of long-range transportation, but it is sufficient for urban and work commuting—which account for most of the driving people do in a year. This does not represent a niche but the largest part of the market.

Since the electricity distribution infrastructure is already in place, cars would simply be recharged at home from a regular power outlet, often taking advantage of cheaper off-peak energy rates at night (assuming supportive regulations). That would change the economics of electricity.

Batteries could be an issue in northern climates. A combination of better insulation and additional infrastructure could help solve the problem. In the Canadian Prairies where winters can be bitter, parking lots are supplied with electrical outlets so that cars can be plugged in even while at work. That could also be part of the solution. The additional infrastructure would represent some expense, but all its components are mass produced, can be quickly installed, and require little maintenance. At this point, the future of the electrical car in northern climates may still depend on technological advances.

With the exception of batteries, electrical technology for cars is low cost because it is already well established and mass produced. Maintenance for motors is also much less expensive than for combustion engines as there are no carburetors, radiators, oil changes, etc. That would also mean a lot less metal and a lower initial price under an environmental taxation system. As electrical cars are not currently mass produced, they are bound to be more expensive than they will be in the future. Their simpler and lighter technology should eventually make them significantly cheaper than their gasoline alternative.

Conservational and environmental cars could be designed and mass-produced relatively quickly if governments and industry cooperated to speed up the process. Regulations could bring in cross-industry uniform chassis for longer lifespans, maximize the use of standard parts, and establish modularity to enhance repairability and allow for easy future fuel-cell conversion. This strategy could bring about a fair amount of synergy, leading to cooperation within industries, reducing risks, and avoiding the loss of investments. Convertibility would also prevent a large amount of resources from being invested in obsolescence.

At the moment, the incentive to move to clean-powered vehicles is still small. The automobile industry has started designing and producing greener models, but there is a limited market for them. The ETS would increase the demand for greener vehicles and usher in new generations of conservational (smaller, less metal intensive, built to last) and environmental (powered by clean and renewable energy) vehicles.

## SPV Transportation: Size Does Matter

We have now progressed from gas guzzlers to convertible electrical cars. The ETS would take us one step further and change the urban environment as we know it.

A four-passenger automobile is not needed to transport only one person. Significant amounts of resources and energy would be saved in designing one-passenger cars for work commutes. The benefits of one-seater vehicles are many: smaller automobiles are more maneuverable in congested traffic, are easier to park, and are generally less costly to drive. Single-passenger vehicles would cost less, use up less non-renewable resources in their manufacturing, and probably be more than twice as energy efficient. Lower weight would also extend their driving ranges.

There is a market for SPVs. As well, there are price, conservational, environmental, and traffic incentives to minimize the size of cars. The new single-passenger vehicles would be not only shorter, allowing twice as many cars in a traffic lane, but also thinner, allowing two vehicles to drive side-by-side within it.

At moderate and high commuting speeds, SPVs would run staggered rather than parallel. They would do so on different sides of the same lane, allowing twice the number of vehicles within a given space while respecting safe driving distances and easing pressures on circulation. As they slow down to approach intersections, stop signs, and red lights, or are caught in traffic jams—where their congestion-reduction ability would matter most—they should be able to run side-by-side within a single lane, a given space packing in as much as three or four times the number of vehicles it currently does.

Single-passenger conservational and environmental cars would nearly quadruple the current traffic capacity of roads. This could eliminate most of the circulation problems that plague commuters on a daily basis in most large cities around the world. Of course, this assumes that we do not increase the total number of cars on the road. The ETS would not lead to that if appropriate tax rates are set. It would make cars generally more expensive, preventing their proliferation.

In fact, if we increase the number of vehicles on the road, the purpose of designing more conservational cars would be defeated and even more non-renewable resources would be consumed. It would also undermine environmental alternatives such as public transit, car pooling, and cycling. It would be up to governments to set taxation levels appropriately to prevent this from happening and achieve what they would like in terms of transportation for the future.

SPVs would yield on their own significant conservational and environmental benefits. Reducing the total number of vehicles at the same time could further increase environmental gains but would call for greater political will and might not be viable for that reason. In the short term, maintaining the status quo in terms of number of cars on the road is probably the best policy as it would allow governments to bring in the ETS with much less resistance on the part of the general public.

Fourth Wave urban environments would be drastically different. SPV transportation would be clean (emission free, smog free), quiet (electrical motors can hardly be heard), and traffic-jam free in most cases.

## Technical Issues Relating to SPVs

There is a number of technical issues relating to SPV transportation. Single-passenger car width would have to be restricted if two of them are to fit within a single lane. Cities may choose to redraw some traffic lanes. Vehicles would have to be properly designed to prevent rollovers upon turning. Architectures would have to include such things as swivel technology and low centers of gravity. Car speed may have to be limited.

## Trends for the Future

ETS-based transportation would call for much higher quality vehicles. They would be built to last for a long time and be more repairable and upgradable. Some parts are already fairly standard in today's cars— wheels, tires, batteries, mufflers, etc.—and things like brakes can be refurbished. The used car industry already exists. The ETS would not create those trends; it would push them much further.

Although advances will continue to be made, we already have the technology necessary for electrical vehicles. What is currently missing is a market. The environmental taxation system proposed here would create one.

How farfetched is all of this? We do not see SPVs on the market at this point in time. However, the size of cars is decreasing, and the automobile industry is increasingly talking about *biocars*, vehicles having parts made with bioplastics and composites produced from soy, wheat, canola, or sugar cane.

Some companies have already developed plant-based polyurethane foam for seat cushions. Volvo claims to use renewable materials in dozens of its car parts. Mercedes S-Class vehicles are also going green, their bio-components claimed to tip the scale at 43 kg per unit (Stauffer, 2008, February 15). SPVs are not here yet, but the automobile industry is definitely moving into greener fields.

In the first half of 2008, the sector saw tremendous changes. The sales of SUVs dropped sharply from one month to the next. Major manufacturing companies shocked analysts by deciding virtually overnight to close several SUV manufacturing plants in the US and Canada. Most automakers started talking about either developing or producing electrical cars and super-efficient vehicles. Some have models already on the market.

## The Ready Market for SPVs

There is an existing market for single-passenger electrical cars: couples that already have two vehicles. They do not need two multi-passenger long-distance fossil fuel cars or Sport Utility Vehicles (SUVs) for simple urban commuting. The second means of transportation—if they really need one—could easily be an SPV.

Currently, most cars making the daily work commute are four-passenger vehicles which convey only one person. Our current way of getting around is highly inefficient not only from a fuel perspective but also from a non-renewable resource point of view. The second family car is a ready market for SPVs.

Other possible buyers for the one-seater electrical vehicle are single people. Instead of purchasing a gasoline car, many of them would choose to buy a lower cost electrical car for commuting to work or school, especially once the ETS makes vehicles pricier. They would rent a gasoline automobile or a hybrid once in a while as necessary, for example, for vacationing. Single people are another ready market for SPVs, at least once they are mass-produced and their price drops.

Remember that the ETS would be revenue neutral, and that while cars would be more expensive, people would have more money to spend.

## Safety for All

The minivan and SUV markets really took off on the safety issue: the bigger, the safer. However, the real question is, safer for whom? Although they offer more protection to their own drivers, they may not be better for the rest of us. Much bigger vehicles are more dangerous to both other drivers and pedestrians, at least in theory. In my view, greater safety for all lies in decreasing the average size of vehicles, not the opposite.

A lack of safety does not stop pedestrians from crossing the streets or other people from riding bicycles and motorbikes—all of whom have virtually no protection in collisions with cars. As such, a market for SPVs will develop regardless of the safety issue.

Smaller cars are often thought to be less safe for their own drivers. This is likely true to some extent. However, they might not necessarily offer much less protection than other vehicles. Design is a large component of safety. Race cars are smaller, yet they provide higher protection than your average automobile. Properly engineered SPVs could offer a reasonable amount of protection. They would also provide much safer city streets for everybody.

## Fast-Tracking SPV Transportation

SPV transportation would bring in vehicles about 65% smaller than current four-passenger cars. That would mean, in theory at least, a 65% conservation of non-renewable resources—metals—a 65% reduction in intermediate chemical use, and at least a doubling of fuel efficiency. Much smaller, resource-efficient, and more repairable one-passenger vehicles would offer a number of cost savings. This should initially translate directly into greater affordability compared to regular cars. Mass production would really be the key to giving SPV transportation a competitive edge and, in doing so, achieving massive conservational, environmental, and traffic benefits.

A concerted national strategy could bring electrical-vehicle transportation into mass production within a couple of years. Governments and automobile manufacturers could get together to design within a year a uniform chassis for a one-passenger car that would be mass produced and used as common architecture for all manufacturers in their first models. Cooperation would lower development costs, enhance modularity and reusability of chassis, and enable mass production even for the first models.

Pooling R&D would force on the industry efficiencies which would be beneficial to both consumers and manufacturers. Furthermore, standardization could also occur internationally. Modular designs would create a platform that is both conservational and environmental and do so on a massive scale worldwide.

ETS taxation on non-renewable resources and stable high oil prices would be the basis that would provide for steady and predictable growth of not only the renewable energy sector but also SPV transportation and green technology in general. Appropriate government support and industry cooperation would eliminate the high development risks and insecure markets for car manufacturers.

The consumer would be happy, the industry would be happy, jobs in the new automobile sector would be preserved, resources would be saved from building much smaller vehicles, and energy efficiency would be improved. Dependency on Middle Eastern oil could quickly decrease.

The demand for electricity—which has the potential for being produced cleanly—would increase. The markets for wind turbines, solar panels, and other renewable energy technologies would take off. Micro-producers would join in the frenzy, selling excess electricity into the grid and improving their own country's balance sheet. The shift to electrical vehicles in urban transportation would decrease the demand for biofuels and lower pressure on the price of food, perhaps averting widespread instability and a world hunger crisis.

SPV transportation would mean massive gains in resource conservation and energy efficiency. Those would occur to a large extent in a market-friendly manner. As necessary, regulations and incentives—including tax breaks and rebates on insurance—could be used initially to enable a quick takeoff for SPV transportation.

## International Markets

### Developed Countries

A huge market for SPVs would be in developed countries where they could provide an alternative to many of the cars on the road today. Together with double-passenger vehicles based on the same conservational and environmental technology (DPVs), they could take over the bulk of the market for cars.

### Developing Countries

Most countries around the world cannot afford a North American style of transportation. Neither can the planet. A proliferation of cars around the globe would be disastrous for everybody. The question of transportation in the developing world is highly complex. Many countries already have a significant part of their transportation based on smaller vehicles: gasoline rickshaws, motorcycles, scooters, etc. Replacing those with cars, even small ones, could do more harm than good.

Remember that the ETS would make vehicles more expensive. As such, we should not witness a proliferation of cars or upgrades to bigger vehicles on its account. Of course, those might occur as a result of economic growth or government policy. That question is discussed below.

A switch from fossil-fuel to electricity-based transportation would be a major improvement for many developing countries. Smog and greenhouse gas emissions would be significantly reduced. The demand for petroleum would decrease, which would serve to keep its price moderate. More and cleaner energy would be produced domestically, generating wealth and creating jobs at home.

A properly implemented ETS scheme would lead to gasoline rickshaws, motorcycles, and scooters being replaced by their electrical equivalents rather than being upgraded to cars. The technology exists today. For example, several companies already make battery-powered scooters for the North American market. Electrical rickshaws are currently being produced for many countries around the world. As such, SPV transportation would very likely be a major improvement over what already exists in most developing countries.

Together, China and India host about one third of the entire world population. Their economies are some of the fastest growing today. Each has a significant lower middle class. Expecting them to decline an improvement in their standard of living may not be realistic. Under the most likely scenario, a significant increase in the number of vehicles in these countries is inevitable. What is for sure is that it will not be our choice to make.

SPVs based on electrical technology would represent a better option for developing countries. Their lower cost, lesser use of metals, easier maintenance, and smaller ongoing energy needs would provide both affordability and environmental benefits. In comparison, hydrogen would require more expensive technology and huge investments in infrastructure. It would simply not be as good for those reasons.

Tata Motors in India is planning to put on the market gasoline sedans that would cost as little as U.S. $2,500 each. In the opinion of many environmentalists, that could spell environmental disaster for the one-billion-people country. The company is also planning to expand into international markets. This could lead to a proliferation of cars around the world, especially in developing countries.

To cater to its own growing middle class, China will either import India's cheap automobiles or follow suit with a similar model. The country will likely also want to take advantage of international markets. This would be disastrous for the planet. SPVs might be a better alternative and would have huge potential sales in both the developed and developing world. So would double-passenger vehicles based on the same technology.

The ready market for SPV transportation is actually really, really big internationally. One questions is, do we want a proliferation of gasoline vehicles in the developing world as it seeks better living standards, or do we want those to be emission-free SPVs? A second one is, who will take advantage of this opportunity first: North America,

Europe, the Far East? A third one would be, once India and China start delivering cheap gasoline cars to the world markets, what are the current leaders in the automobile sector (North America, Europe, Japan, etc.) going to produce and sell, if not the next generation of cars?

# 15. The Global Environmental Accord: Beyond Kyoto

International accords could significantly boost worldwide resource conservation and environmental protection by reducing competitiveness issues relating to the implementation at the national level of the taxation system proposed in this book. The ETS should be the object of an international agreement—which will be referred to here as the Global Environmental Accord (GEA)—just as we did with the Kyoto Protocol. First, let us take a look at the latter to see what we can learn from it.

The Kyoto Protocol is perhaps the world's greatest cooperative effort so far in addressing environmental problems. However, it fails in many ways. Firstly, it is a regulation-based approach that is fairly bureaucratic. Secondly, it attempts to address essentially only one issue: greenhouse gases. It does nothing for the conservation of non-renewable resources (except indirectly for fossil fuels) or other environmental problems such as contaminants. Thirdly, the intent of the accord is not to stop global warming but only to slow it down. This is admitting failure from the start.

There are better ways than regulations to achieve the ends of Kyoto. A market-based approach would be simpler, have lower administrative costs, and could gain broader support than the accord did.

## The Kyoto Protocol

The Kyoto Protocol is being renegotiated at the moment. It is difficult to know where it is at exactly or who is party to it or not. Broadly speaking, its first stage proposed to have countries reduce their emissions of greenhouse gases for the period of 2008 to 2012. Different groups of countries have different reduction targets expressed as a percentage of their 1990 emissions levels. For example, Canada and the US had to reduce them from 2008 to 2012 to about 6% or 7% lower than they were in 1990.

One of the issues raised with Kyoto is that countries that had underperforming economies in 1990 have much stiffer standards to meet, their emissions having been lower than usual in that specific year. Another issue is national growth levels since 1990. Countries that had stagnating economies during the 1990s are at a significant advantage compared to others, their emissions having increased less than the rest of the world. As well, countries that have seen significant increases in population and economic growth during that period have much further to go to reach Kyoto targets.

Governments that fail to reach the agreed upon targets in 2012 would have to purchase emission credits from outside, which in itself is generally viewed as an efficient way of doing it. That would probably entail the setting up of an international trading system of some kind. Canada's strategy involves caps on total emissions for industry sectors, internal credit trading mechanisms, and price ceilings for emission-reduction costs for companies (Doucet, 2004). This may be somewhat different at this point in time as policies change with governments.

The whole system is likely to be fairly complex and a significant burden to the industry. Kyoto's structure is essentially regulatory, cumbersome, and bureaucratic. The Global Environmental Accord could broaden it and bring in more efficient mechanisms to reach its targets.

## The Global Environmental Accord

The Global Environmental Accord would involve three different groups of countries divided according to income levels: high, middle, and low. Of course, this could be expanded as necessary. In some cases, policies would be applicable across the board to all the different groups. In others, different levels would be applied according to purpose and ability. The comprehensive accord would comprise three components: non-renewable mineral resources (except energy), fossil fuels, and environmental standards. Initially, the main intent would be to establish a flexible framework that is acceptable to the largest number of countries.

### The Non-Renewable Resource Component

The non-renewable resource component of the Global Environmental Accord would be crucial for not only beginning the global conservation of resources but also supporting individual countries' national implementation of their own environmental taxation system. As already expressed, without multinational agreements a certain amount of conservation could be achieved, but progress would eventually reach a ceiling as any country opting for a resource taxation strategy would decrease its international competitiveness.

The Global Environmental Accord would establish taxation standards for non-renewable resources, for example, 25%, 35% , 50%, or whatever would be deemed appropriate. Initially, these could be set lower in order to establish a broad-based common framework. They could be raised later to desired levels as economies begin adjusting to the new green economic environment.

The rates would be fixed. Countries would not try to meet a target price as in the case of oil. The levy would be charged on both domestic supplies and imports. This would lead to conservation at home but would not hurt local economies as importing countries would themselves collect the levy. Consumers would get income and retail tax rebates as per the revenue neutrality principle.

As such, regular international prices would remain in effect for importers, and the system would not penalize resource-poor or developing countries as long as they participate in the accord. Whenever sold to non-GEA members, raw materials would be charged a levy as they are exported.

Mineral-rich states in both the developed and developing world would benefit from this component. However, poorer countries that do not have substantial mineral resources could be hit hard, paying higher prices for imported finished goods with high metal content. Foreign aid could be redirected to them to compensate for this.

The new scheme would, for example, work like this. High-income countries would have to enable full taxation. Middle ones would have more flexibility, being expected to tax to a minimum of 80% of the established standard. This would promote exports for them and economic growth. Low-income countries would have full flexibility. This would let them choose what is best for their particular situation. Of course, nothing would stop them from still choosing to meet the full standard as it would maximize conservation.

Higher world prices for non-renewable resources would reduce demand and global depletion while offering middle and low-income countries options ranging from implementing the full standard to exercising varying degrees of flexibility in taxation in order to boost their developing economies.

Because the ETS tax would be collected as tariff by importers, the approach would be revenue neutral for individual countries and the money raised would not be lost. It would be plowed back into local economies through reduced income and retail taxation. Under the GEA, countries that are resource poor would not be disadvantaged, and those having plentiful supplies of minerals would not enrich themselves at the expense of others as is the case with OPEC.

As the price of rice skyrockets, some countries are talking about forming a cartel for the commodity. The GEA would give us a mechanism to prevent a future of cartelization in which one commodity after another would become the object of speculation. The OPEC way may not be what we want as a future for our children and people around the world.

### The Fossil Fuel Component

For the sake of simplicity, the following will focus on oil and global warming—the one issue addressed by the Kyoto Protocol—but the GEA would also involve other fossil fuels.

Under the Global Environmental Accord, oil would be the object of domestic and import taxes that would raise its cost within individual countries to a certain level. Initial price targets could be chosen so as to approximately achieve the global emission reductions already decided under the Kyoto Protocol, making everything simpler and easier to agree on. The ETS fossil fuel strategy would call for different price levels for each group of countries in order to maintain the status quo of Kyoto in this respect.

As in the non-renewable resource strategy, high-income countries would be expected to meet the full target whereas other groups would be given greater and full flexibility. Higher prices would lead to conservation and promote the growth of renewable energy sectors. For the US and other net importers of fossil fuels, that would mean increased wealth as money would stop hemorrhaging out of their own economies.

After setting the initial target for oil, there would essentially only be a minimal amount of bureaucracy and no emission credit systems to establish. Enforcement throughout the entire business sector would boil down to regular tax collection from a handful of corporations. All a government would have to do is keep track of the international price of oil and establish the appropriate levy to be applied to both local and imported sources at the wholesale level. The tax would be revised periodically as the international cost of oil rises and falls.

The industry would not face a regulation nightmare or need to spend time understanding and implementing arbitrary rules. No rigid standards would be applied throughout an entire sector. Businesses would face simple input-price decisions for which they already have full expertise. The approach would be dynamic and flexible. Price levels in the accord could be renegotiated every few years in order to meet new emission targets. The tax would not be charged on oil exports so that international prices continue to be set through markets. Decreasing world demand would moderate increases in the price of petroleum.

A full ETS energy strategy would also tax other fossil fuels, such as coal and natural gas, not only because they are depletable and produce greenhouse gases but also to prevent a shift from one source of emissions to another.

### *The Environmental Standards Component*

The GEA's environmental standards component would serve to integrate and bring into a common framework existing regulations on pollutants and contaminants. The component would mitigate the negative effects of the stricter environmental regulations of some countries on their international competitiveness, promoting better national as well as worldwide standards.

Regulatory limits and bans would play an important role in this component. However, the accord would still be primarily market-based, relying on taxation to deter the use of a range of pollutants and contaminants. The component would integrate existing environmental standards on highly toxic compounds and any future developments in this respect as determined by the international community. Other chemicals would face taxation based on their toxicity, bio-accumulation, and potential for cumulative damage to the environment.

Operating on a formula similar to that of the non-renewable resource component—different levels of flexibility applying to group of countries—the market-based approach would increase environmental standards across the board. Wealthy regions would have to meet full requirements. Middle-income countries could opt for softer targets as under the other two components. The poorest would have to adhere to a less stringent standard although they could still choose to do better and enjoy a cleaner environment.

Negotiations could lead governments choosing to raise requirements for all countries proportionally, making the world a better place for everybody while maintaining the status quo in terms of international competitiveness. Another option would be to increase them for the top two groups only, yielding positive results for them in the form of a better environment and, by the same token, increasing the international competitiveness and potential for economic growth of poorer countries.

## The New Environmental Politics

The GEA would not be just another environmental accord. Its politics would be very different from those of the Kyoto Protocol—which has suffered from a lack of support on the part of countries like the US, China, and India. These three alone represent about one third of the entire world population.

Participation to the Kyoto Accord is essentially voluntary. Nobody would be forced to join the GEA either. However, importers of raw materials would find themselves dragged into conservation if producer countries decided to implement the ETS.

Canada is a net exporter of raw materials. The US is the opposite. Suppose that only the Canadian government decided to join the Global Environmental Accord and tax minerals. The US would be paying a higher price for its imported raw materials because of the GEA export tax to non-members. It would therefore have an incentive to conserve. Whether Americans join a global accord or not, they would be caught in its wake.

With respect to the above, there would be primarily two effects to be concerned with. In response to higher import prices, the raw materials sector in the US would become more profitable, leading to an increase in domestic production and the creation of jobs. As raw materials are more costly to produce in the US, consumer goods made with domestic supplies would be more expensive, resulting in a certain amount of conservation.

The important conclusion here is that the implementation of an ETS type of approach in countries that export minerals would result in worldwide conservation benefits regardless of the existence of an international accord on the issue.

The second effect would be a potential shift of wealth to countries that are net exporters of raw materials as is currently occurring with oil. Suppose that Canada produces steel for a basic cost of $1,000 per ton and adds an ETS levy of 20% on top, yielding a total export price of $1,200. It would collect $200 per ton of steel in taxes from Americans. As such, some wealth would be shifted north.

Membership in the Global Environmental Accord would prevent that from happening. Without the GEA, the ETS would lead to the enrichment of the countries having the most plentiful reserves of minerals on the planet. Raw materials would be taxed as exports, and producing countries would reap windfall profits as importers pay much higher international prices for the commodities.

Under the GEA, the import tariff paid by Americans would remain in the US instead of being shifted north. Of course, this assumes that the country joins the accord. American manufacturers would make goods with steel that is more expensive—leading to conservation— but at the same price as the one used by their Canadian counterparts, ensuring that the competitiveness between them remains unchanged.

Producer countries could also choose to not fully tax their exports as an alternative means of handling the issue.

In the end, once mineral exporters decide to implement the resource conservation component of the ETS, the rest of the world would have no choice but to follow suit. Joining the GEA would only benefit them. As such, participation in the accord would be much less of a problem than it has been for the Kyoto Protocol.

## A View of the Future

In most respects, the Global Environmental Accord would be more flexible and scalable than a regulatory equivalent. It would provide an international framework for the world to begin the conservation of non-renewable resources, address global warming problems, and increase environmental standards across the planet.

The Global Environmental Accord would result in many countries decreasing their dependency on oil. Their renewable energy sector would take off and not look back. Resource conservation would soften the impact of scarcity and delay potential crises. The use of contaminants and pollutants would progressively decrease from the continuous incentive to do better provided by a market-based approach.

The alternative course is the current one. Increasingly higher international oil prices will lower our standards of living. Other mineral resources will likely follow a very similar pattern, only a bit further down the road. It is a very dark future: resource depletion and excessive prices, decreasing real incomes, environmental contamination worse than it already is, and global instability growing from the power shifts associated with scarcity. That future is not that far away. It is probably our children's and grandchildren's!

# 16. The Environmental Revolution

The ETS provides a mechanism that could bring about the large-scale changes that we need to trigger an environmental revolution. It could do so because it is revenue neutral and would generally not cost the taxpayer a cent. The massive need for funding required by other proposed environmental strategies is primarily why we have failed so far in making serious environmental progress.

Our habits would have to change and become greener, but we would generally enjoy a standard of living similar to the one we are used to. An environmental revolution in the 21$^{st}$ century would require not only an effective strategy—the ETS—but also a number of key ingredients: fundamental legitimacy, social support, and timing. In addition, we would also need a plan of action.

## Fundamental Legitimacy

The fight for the environment certainly has undeniable and fundamental legitimacy: ending the pillaging of non-renewable resources and protecting the environment for ourselves and posterity. The gouging of resources such as metals is a one-way street. There is no second chance. Being down to counting the number of toxic compounds in the bodies of unborn babies—as research currently does—should be a clear message that it is time to act.

## The Demographics

Politics is a numbers' game. There would be two main demographics involved in an environmental revolution: baby boomers and post-boomers. The former are currently at the helm and will be for some time. They are the ones who have the political power to bring about an environmental revolution.

Post-boomers are the new wave of consumers. They partly keep companies in business today and will be the ones patronizing them tomorrow. They have immense economic clout.

## The Timing

Just a decade ago, people were buying SUVs and gas guzzlers as if there were no tomorrow. With the relentless escalation of oil prices and the building of a consensus about the fact that these will not go back down, people are slowly waking up to the possibility that the fossil-fuel honeymoon might just be over. We need to get real about increasing efficiency and shifting to green energy.

Just a decade ago, the average person was not too concerned about global warming. Since the flooding of New Orleans, attitudes have changed dramatically. People have been dying for a long time as a result of our lack of concern for the environment, but New Orleans crystallized the issue. We are now increasingly realizing that contamination and the *carbon machine*—all the activities contributing to global warming—will not stop on a dime and are even very difficult to slow down.

The fossil fuel and greenhouse gas issues have broken through to popular awareness. They are the object of a major international agreement (the Kyoto Accord) and appear on the news media on a daily basis. Unfortunately, contaminants seem to have taken a back-seat, and ore depletion concerns have not received as much publicity. Most people are not aware that they are also in dire need of being addressed.

The timing for a revolution is good for two reasons. Firstly, there is a much greater awareness about environmental issues than there was before, hence more political support. Secondly, there is an urgency to act with respect to not only global warming but also contaminants and the depletion of resources.

We already have an environmental strategy—the ETS; now we need a plan of action.

## The Plan

We have all the key ingredients for an environmental revolution: fundamental legitimacy, the demographics, and the timing.

A lot of research has gone into this book. It was designed as an educational and promotional tool. Much effort was made to write it in a language and format that most would find easy and interesting to read so as to reach a large number of people and have a maximum political impact.

Great care was taken to ensure that the book cover most of the arguments that might come up against environmental taxation. As such, anyone reading it would have all of the important facts and would not easily be fooled by false claims made by detractors or opponents to environmental change.

As such, a publicity campaign to promote an ETS type of approach could be achieved easily by simply getting people to read this book.

### *Your Part*

Your helping to publicize this book would achieve two things: raise awareness of the ETS and support the work I did and will continue to do.

There are a number of ways in which this can be accomplished. Below are several options, but be sure to check the Waves of the Future website (http://wavesofthefuture.net) for updates and other possible strategies. There will be projects in which you can partici-pate as well as online tools, a blog, and a forum. If you have any suggestions, you will also be able to make them through the website.

### *Word of Mouth*

Obviously, the easiest way to help with this book is to suggest it as reading to anyone you know. Getting bits and pieces from the media is not going to be enough. People really need to be aware of all of the arguments. Political support will be much stronger and more effective if they do.

You can also publicize the book on any available bulletin board. There are some in shopping centers, grocery stores, condominium projects, community centers, at work, etc. Most schools, colleges, and universities have several of them. You can stake out three or four of them and make sure that they regularly display a note about the book.

You can also try to involve any group that you belong to, environmental, political, or other. Make sure that you give people the website address so that they can find other ways to participate and the most up-to-date information.

## Online Promotion

One of today's most effective ways to advertise and promote something is the Internet. You can use every available opportunity that you have in this respect: blogs, forums, your own website, etc. You can add comments in your MySpace and Facebook pages or on any other network of the kind. You can also make a YouTube video.

Make sure that you always provide the website address for the book so that people are able to find more information about it. Use a search engine to find environmental blogs, forums, discussion groups, etc. Do some of these things on a regular basis, for example, once a month. Doing it one time only will not be enough.

An obvious means of promotion is to write a review of the book on websites—such as Amazon.com, Lulu.com, and others where the book is sold or environmental issues are discussed. It does not have to be long but could make a lot of difference. If you are not inclined to writing, just rate it. This is very useful to readers when it comes to purchasing a book that they have not already heard about.

Another useful approach is linking to the Waves of the Future website. Doing this makes it more visible to search engines. It is not simply typing the website address. In some cases, the software or page that you use will create a clickable link automatically. If it does not, the correct HTML coding is:

`<a href="http://wavesofthefuture.net/">wavesofthefuture.net</a>`

Just type it exactly as shown above. When you upload or send the information, a blue clickable link should display with the appropriate name. If that fails, check the Waves of the Future website or email if the correct form is not already there.

Along the same line, the Internet now offers a multiplicity of means which you can easily use on your own web page for promotion—buttons, widgets, mini-storefronts for FaceBook and MySpace pages, book cover pictures, etc. Codes for these will be available at the website.

### The News Media

You can bring the book to the attention of the news media. Write to book-review programs on television, columnists, or environmental correspondents. Give newspapers your opinions through their *Letters to the Editor* section. Anyone can do that. Many publications now accept submissions via email.

### Best Places to Buy the Book

Royalties to authors can vary hugely depending on where you purchase a book. In this particular case, the best places to buy it in order of preference in the US are CreateSpace.com (an Amazon.com company), Lulu.com, and Amazon.com itself (the original imprint). They all ship worldwide.

As I write, the best place for international customers (except Canadians) is Lulu.com. It has low shipping fees (for example, £3.5/€4.99 or less for the EU; $6.50 to Sydney, Australia) and is associated with local printers in the UK and Spain. That would probably mean intra-EU shipping. Verify this with Lulu to be sure.

In Canada, the best place to purchase is from the Waves of the Future website (low shipping fees). Printing is done within the country, and items are sent out from a local warehouse. Autographed copies should be available from there for Canadian, U.S., and international customers.

## Governments and Corporations

Governments and corporations may be able to help. This is a major project that would benefit society as a whole. Countries may want to participate or support this in one way or another.

Supporting the environment has become excellent public relations (PR) for corporations. Here is a chance for them to do something significant. Suggest it to your employer. The website will provide information on how they can help. Suggestions are also welcome.

Green businesses would hugely profit from the ETS. Supporting this book is probably one of the best investments they could ever make to increase their markets as well as speed up the development of the entire sector. Suggest it to them.

## The E-Book Project

The most obvious way to get involved for governments, corporations, and well-off individuals is to participate in the *E-Book Project*. The most important way authors make money is by selling actual books. With today's Internet piracy problem, illegal digital copies of one's work are too easily made and distributed, deterring many authors from offering electronic versions of what they do.

E-books are much cheaper to produce and can generally be sold at a much lower price to consumers. An electronic version of *The 21st Century Environmental Revolution* could be made available cheaply or even for free with donations from or the sponsorship of corporations, organizations, and individuals. See the Waves of the Future website for details.

Most people in the developed world can easily afford a book at its regular price. That is not necessarily the case elsewhere. It would be very important for all developing countries to join in the fight for the environment and participate in a global accord. Free digital copies of this book could make a lot of difference in terms of international promotion and political support for the ETS and the GEA.

The *E-Book Project* would benefit everybody. Readers and fans would access the book for free, which would help promote major changes for the environment. Sponsors would get the satisfaction of contributing to a worthy cause. Their donations would go towards meeting their social responsibility goals.

Millionaires and billionaires could sponsor free e-books for everybody without even blinking: from owners of major corporations to investors, celebrities, and musicians. You can write to them.

There are thousands of professionals in North America alone, be they medical doctors, professors, engineers, architects, lawyers, computer programmers, etc. It would not take substantial donations from too many of them to reach the goal of free e-books for everybody.

And, there is the rest of us. The current population in Canada, the US, and Europe is over one billion. If only one out of every 10,000 people in those wealthy regions of the world gave $10.00 towards the *E-Book Project*, more than enough money would be raised. Ten dollars is the equivalent of a couple of morning stops at the coffee shop or breakfasts in a fast-food place. Can you spare this much for the environment and you children's future?

### Reading Lists

You can add the book to online reading lists that you know of or inform me about them. You can write to people hosting their own recommended lists and suggest that it be added to them. Some are very popular and would provide a huge amount of publicity for the book.

### You

Obviously, *you* are the key to all of this. If *you* assume that *somebody else* will do it, the environmental revolution could be dead before it even starts.

## Consumer Power

Consumers have an enormous amount of leverage. Everything we buy is a vote we cast. Every time we purchase something, we either reward environmental practices or punish them. We can buy green or purchase wasteful or excessively packaged goods and reward companies that do not care about and destroy the environment. As consumers become more and more aware of their power, attitudes in the business world will have to change. It is up to us. Some of this has already begun to happen, but there is a long way to go.

Our consumption patterns will define our future and that of our children and grandchildren. *Conscious consumerism* can go a long way in making tomorrow's world better. It is one of the most powerful weapons we have today. The ETS would bring in the profound changes that we need for the environment and magnify consumers' power by making green products cheaper and promoting a natural switch to their markets. With it, consumers will be able to choose the kind of world that they want to live in.

## The Industry as Partner

The industry can be a powerful ally in implementing the ETS. Many sectors would benefit directly from the emerging green markets, but there would also be many promotional opportunities for other companies if they join the fight for a better tomorrow. For example, some chemical producers have already shifted a fair amount of their R&D towards environmental substitutes and rake in a lot of free publicity just for doing the right thing.

## Future of Scarcity or Greener Society

According to most estimates, oil reserves have already peaked or are expected to do so within the next 20 years. Other mineral resources will follow a similar pattern of increased costs and scarcity as world economies continue to grow. We will start being hit with shortages, prices will rise, and we will face crises as we have with oil. That may be only a few years or decades away.

The US is especially at risk for resource shortages because it is, and has been for a long time, a large manufacturer and producer of goods. It imports minerals from Canada and many other countries around the world to feed its growing industry. The US could face very difficult times ahead. Without the ETS and resource conservation, it will likely have to contend with more crises in the future than it has in the past, resulting in increasing poverty. This is true of many other nations as well.

Both advanced and developing countries have every reason to implement and support the accord to ensure their continued economic success and that of their children. The GEA would also prevent the cartelization of one commodity after another. The alternative is the OPEC way and the continuation on this destructive course towards scarcity, crises, and geopolitical turmoil.

## The New Landscape

What would the world resemble under the ETS? What would the green economic environment imply for all of us? Countries would get greener and greener year after year. There would be much less pollution in cities thanks to hydrogen and battery-powered electrical vehicles. SPVs would make most traffic jams a thing of the past, assuming countries opt for non-proliferation policies. Streets would be safer and much less noisy for everybody. Public transportation systems would reduce traffic and be more heavily supported by governments.

Consumers would retain their existing purchasing power, but prices would change to reflect a greener economic incentive structure. Many items would be more expensive, but there would also be more money to spend. Goods would be kept longer and repaired more extensively.

Because of the increase in the price of minerals, consumption would shift towards services and green sectors, promoting growth in various industries. Restaurants, bars, theaters, spas, sports and health centers, personal care companies, etc. would see their business grow. The music, video, and film industries would also do well.

Many pollutants (overt and hidden) would be replaced by greener substitutes. Everything we buy (household cleaners, soaps, etc.) would become less harmful to the environment. The waste disposal industry would change dramatically, transforming itself into a source of renewable and non-renewable materials for the industry. Packaging would decrease and become greener and often reusable. We would recycle much more intensively than we currently do.

The green economic environment would reward people for doing the right thing instead of punishing them as is currently the case.

## *Energy and a Thriving Agricultural Sector*

Countries would produce a lot more of their own energy instead of relying on imports. It would come from various and diversified local sources. This would significantly change balance sheets for oil importing countries, keeping money at home.

Agriculture would become much more lucrative as the demand for its products would increase significantly. As a result, new investment would flow into the industry and serve to boost production.

Governments will likely have to regulate how much good agricultural land can be dedicated to the production of energy. Diversified domestic energy policies would lead to the production of biofuels from non-food crops grown on unusable land, refuse, and industrial wastes and byproducts. Some energy could come from food crops to eliminate the excess production that often results in depressed agricultural prices. The need for subsidies would disappear.

The ETS would lead to a slowdown in the increase of the price of oil, which would have a positive effect on the cost of food. In addition, the system would favor green goods by not taxing them. This would include food. Biofuels could be produced without having to compete for land with edible crops.

## *The New Law of the Land*

The ETS would transform the world in which we live. The incentives to destroy the environment and waste natural resources would be gone and replaced with new ones that would do the exact opposite. Finding ways to improve the environment and stretch out resources—as opposed to destroying them—would become very profitable. The new structure would reshape not only consumer markets but also the industry, which would work with the environment rather than against it.

The companies that produce chemicals would redirect their research and development towards green alternatives to the scores of toxic compounds they themselves invented only a few decades earlier. The clean up of the environment and of industrial processes would begin and become an emerging market.

The future would see better living conditions in cities and a thriving (and unsubsidized) agricultural sector. The business environment would get greener and cleaner. We would buy fewer material goods and consume more services and environment-friendly products. The price of food would drop, and the energy crisis would subside.

# *Conclusion: Getting It Together*

Broadly speaking, there are two main ways to address environmental problems. We could reduce consumption or adopt a *greening strategy*—making what we produce and consume more environment friendly. Both can be equally effective.

The problem with decreasing total consumption is that it would call for a reduction in economic growth, which would mean unemployment and be unpopular politically. Few would support that at the moment, especially in a context where many countries are still poor. Even the richest parts of the world still put people under bridges to sleep at night.

The ETS would not reduce consumption but make the products we buy greener. It would shift markets away from goods that are unenvironmental. We do not necessarily need to reduce the consumption of services or products that are green. As such, economies could continue to grow and eventually do so in a sustainable way in the future.

Whether they will or not would depend on the strength of implementation of the ETS and on how much greening occurs as a result. Current efforts are not enough. Low ETS levels would not allow for sustainable growth either. Appropriate taxation rates would have to be implemented.

Population growth would increase total consumption. If that problem is not addressed, even the ETS may not be able to cope in the future. The equation is simple: more people equals more consumption and pressure on resources. Even current world population levels would be impossible to sustain if everybody consumed like North Americans.

The ETS will not work miracles. It will only be able to provide for sustainable growth and the improvement of standards of living for the poor in the future if the world population is brought under control and the number of people on the planet eventually begins to drop.

There is no question that we need an environmental revolution. Incremental change does not cut it. There is no question that the sooner we start, and the faster we can make it happen, the better. The two trillion dollars' worth of incentive created by the shift to dual-purpose taxation would be a very powerful engine that could bring about an environmental revolution early in the 21$^{st}$ century and throw us into the very midst of the Fourth Wave.

It is really up to us. Baby boomers and post-boomers must join hands in the fight for their own as well as their children's future. This world is ours to make. We have the means, the numbers, and the engine of change.

## The New Lifestyles

The ETS would redefine the world we live in. It would change the very structure in which businesses operate. Everybody would have a vested interest in shifting to greener habits, pursuing greener ways, and developing greener products.

The scientific breakthroughs and economic performance of the 20$^{th}$ century have led us to expect unlimited growth and a perpetual bettering of lifestyles. However, part of the prodigious productivity we have experienced so far was derived from fossil fuels, a bountiful and cheap source of energy. Another part resulted from our reckless depletion of non-renewable resources. Those were and are used as if they are available in infinite supply.

The era of plenty is coming to an end in both cases. The oil crises we have been experiencing are only the first signs of what awaits us down the road. When scarcity hits other minerals, there will be no second chance.

We have made little headway into solving many of the problems of the last century. It is time to challenge ourselves.

## Final Words

With the revenue-neutral ETS, we no longer have an excuse not to move ahead with an aggressive conservation and environmental agenda. It is now up to us—baby boomers, post-boomers, voters, and consumers—to choose the world we want to live in, to choose our legacy. It is now up to us to bring about the 21st century environmental revolution. It will not just happen on its own.

It is time for change.

**The next edition of this book will likely not be advertised. It will only be made available through very few online retailers. Check the Waves of the Future website for availability.**

# Book II of this series is in the works.
**Check the website below for release dates and places to buy.**

**http://wavesofthefuture.net/**

# *Reference List*

Babcock Gove, P. & the Mirriam-Webster editorial staff (Eds.). (1966). *Webster's third new international dictionary of the English language, unabridged.* Springfield: Merriam.

Bailey, I. (2002). European environmental taxes and charges: Economic theory and policy practice. *Applied Geography, 22(3)*, 235-251.

Baillargeon, J. (2002). *Market and society: An introduction to economics* (M. Galan, Trans.). Halifax, N.S.: Fernwood Publishing.

Beers, B.F. (1986). *World history: Patterns of civilization.* Englewood Cliffs: Prentice-Hall Inc.

Bruvoll, A. (1998). Taxing virgin materials: an approach to waste problems. *Resources, Conservation and Recycling 22(1-2)*, 15-29.

Common, D. (2008, January 25). *Putting the brakes on ethanol.* Retrieved January 26, 2008, from CBC News: Reports from abroad: David Common Web site: http://www.cbc.ca/news/reportsfromabroad/common/20080125.html

*Contemporary world atlas.* (1988). Chicago: Rand McNally and Company.

Council for a Livable World. (n.d.). *U.S. military budget tops rest of world by far.* Retrieved June 05, 2004, from http://www.clw.org/milspend/ushighestbudget.html.

Diamond, J. (2005). *Collapse: How societies choose to fail or succeed.* New York: Viking, Penguin Group.

Doucet, J.A. (2004). *The Kyoto conundrum: Why abandoning the protocol's targets in favour of a more sustainable plan may be best for Canada and the world.* Ottawa: C.D. Howe Institute.

Dryzek, J. S. (2005). *The politics of the earth: Environmental discourses* (2nd ed.). Oxford, New York: Oxford University Press.

Eberstadt, N. (2004). Population, resources, and the quest to "stabilize human population": Myths and realities. In Huggins, L.E. & Skandera, H. (Eds.), *Population puzzle: Boom or bust?* (pp. 49-65). Stanford, CA: Hoover Institution Press.

Ehrlich, P.R. & Ehrlich, A.H. (2004). *One with Nineveh: Politics, consumption, and the human future*. Washington: Island Press, Shearwater Books.

Geller, H. (2003). *Energy revolution: Policies for a sustainable future*. Washington, DC: Island Press.

Hydrogen fuel cells: Clean but costly. (2004, October). *Consumer Reports Canada, 69(10)*, 18.

*International financial statistics yearbook*. (2003). [Washington]:International Monetary Fund.

Isidore, C. (2005, November 9). *Big oil CEOs under fire in Congress*. Retrieved November 9, 2005, from CNN/Money Web site: http://money.cnn.com/2005/11/09/news/economy/oil_hearing/index.htm?cnn=yes

Jackson, T. (2000). The employment and productivity effects of environmental taxation: Additional dividends or added distractions? *Journal of Environmental Planning and Management 43(3)*, 389-406.

Landry, P. (2004, February). *Nicolas Copernicus (1473-1543)*. Retrieved November 4, 2005, from Blupete Web site: http://www.blupete.com/Literature/Biographies/Science/Copernicus.htm

Lean, G. (2004, August 1). *Oceans turn to acid as they absorb global pollution*. Retrieved August 03, 2004, from HealthWorld Online, Healthy News Web site:http://www.healthy.net/scr/news.asp?Id=9602.

Lunder, S. & Sharp, R. (2003, September 23). *Study finds record high levels of toxic fire retardants in breast milk from American mothers: Executive summary*. Retrieved November, 11, 2005, from Environmental Working Group Web site: http://www.ewg.org/reports/mothersmilk/es.php

Lutz, W., Sanderson, W.C. & Scherbov, S. (Eds.). (2004). *The end of world population growth in the 21st century: New challenges for human capital formation and sustainable development*. London, Sterling, VA: Earthscan.

Markowitz, G.E. (2002). *Deceit and denial: The deadly politics of industrial pollution*. Berkeley: University of California Press.

Marx, K. (1957). *Das kapital : Kritik der politischen ökonomie*. Stuttgart: Kröner Verlag.

McIntosh, M. (2000). Globalisation and social responsibility: Issues in corporate citizenship. In Warhurst, A. (Ed.), *Towards a collaborative environment research agenda: Challenges for business and society* (pp. 42-57). New York: St. Martin's Press.

Nadakavukaren, A. (2000). *Our global environment: A health perspective* (5th ed.). Prospect Heights, IL: Waveland Press, Inc.

Pneumaticos, S. (2003). *Renewable energy in Canada. Status report 2002.* Ottawa: Natural Resources Canada.

Pinderhughes, R. (2004). *Alternative urban futures: Planning for sustainable development in cities throughout the world.* Lanham, MD: Rowman & Littlefield Publishers, Inc.

Revenue Canada. (2002). Final basic table 1—sample data, in *Income Statistics.* Ottawa: Author.

*Rocket fuel chemical in California cow milk.* (2004, June 22). Retrieved June 22, 2004, from CNN.com, Health Web site: http://us.cnn.com/2004/HEALTH/06/22/milk.chemical.ap/index.html.

Statistics Canada. (2004, March 11). *Employment by industry and sex* [Data file]. Retrieved March 23, 2004, from http://80-www-statcan-a.proxy1.lib.umanitoba.ca/english/Pgdb/labor10b.htm.

Stauffer, J. (2008, February 15). *Biocars: Your car may soon be more vegetable than mineral.* Retrieved February 16, 2008, from CBC News In Depth Web site: http://www.cbc.ca/news/background/tech/science/biocars.html

Toffler, A. (1970). *Future shock.* New York : Random House.

Toffler, A. (1980). *The third wave.* New York: W. Morrow.

*A unique organisation.* (n.d.). Retrieved May 1, 2004, from Scott Bader Commonwealth Web site: http://www.scottbader.com/pub.nsf.

US Department of Commerce. (2002). *Statistical abstract of the United States.* Washington, DC: Bureau of the Census.

Verbeke, A. and Coeck, C. (1997). Environmental taxation: A green stick or a green carrot for corporate social performance? *Managerial and Decision Economics, 18(6),* 507-516.